William Edward Langdon

The Application of Electricity to Railway Working

William Edward Langdon
The Application of Electricity to Railway Working
ISBN/EAN: 9783743400672
Manufactured in Europe, USA, Canada, Australia, Japa
Cover: Foto ©berggeist007 / pixelio.de

Manufactured and distributed by brebook publishing software (www.brebook.com)

William Edward Langdon

The Application of Electricity to Railway Working

PREFACE.

It is now some four years since the Author undertook the revision of his previous work bearing the same title as that to which these remarks form a preface. Circumstances beyond his control, unhappily, on more than one occasion compelled him to lay it on one side, and it is only of late that he has been able to take up his task, and complete the pledge given to those friends who so frequently honoured him by their request for a reproduction of the book.

The progress which has attended the application of electricity to railway working since the issue of his previous treatise has been such as to call for a completely new work. To its production the Author has bent his mind with the result now before the reader. In doing so he has not departed from the principles underlying his previous work. Much new matter has been added, but in dealing with it he has assumed that for the purpose in view the electrical energy is there. The object of the book is not to treat of the development of Electricity, but to illustrate and explain the manner in which it may, with advantage, be applied to railway working.

Probably within no similar period has the railway telegraph service experienced so rapid an extension as during vi *Application of Electricity to Railway Working.*

the past twenty years. Within that period the mileage of wire solely devoted to the requirements of the railways of Great Britain has expanded by some 40,000 miles, and the number of instruments has been increased by no less than 80,000; resulting in a present total exceeding 100,000 miles of wire, and some 140,000 instruments.

Nor does this entirely dispose of the large telegraphic interest which attaches to the railway service, for to these results should be added some 70,000 miles of wire, the property of the State, which is, at the instance of the Postmaster-General, maintained by the various railway companies.

The fact that the entire railway system of the United Kingdom is now worked under the Block system will, to some extent, account for this large expansion. With what beneficial results the enormous expenditure which must have been incurred by the railway companies in the establishment of this means for the further protection of the travelling public has been attended, the Board of Trade Returns amply testify.

When we realise the fact that the number of messages emanating with one of the larger English companies—that to which the Author is attached—amounted, during the year 1895, to close upon 5,000,000, and that the number of train reports exceeded 8,000,000, we are at no loss to recognise the importance of, and the need which has brought into existence, so extensive a branch of the railway industry. It is, perhaps, interesting in this detail to here note that, assuming the telegraphs of other railways are employed to a similar extent, the number of messages emanating with the railways, quite independent of train reports, approximates closely to

Preface. vii

the telegraphic business transacted by the Postmaster-General for the entire commercial community of the country. The Midland Railway embraces something less than one-sixteenth of the entire length of railway line open for passenger traffic in the British Islands. 5,000,000 × 15·5, will give us a total of 77,500,000, as against 78,839,610, the number of telegrams dealt with by the British Postal Telegraph Department during the year 1895–6.

Electricity has for many years been referred to as in its infancy. The past twenty years has witnessed so many important advances in its application to the needs of every-day life, that it would be rash to assume such is not still its position. The fulfilment of to-day but tempts us to look for still further development on the morrow! To whatever purpose mechanical agency is applied, to whatever purpose it is applicable, there electrical energy will find an economical and a useful field.

That the claims of commerce, competition, and the conveniences of life, will lead to continued expansion and amplification of the several branches of Telegraphy and Telephony, we may learn by reference to any one of the Postmaster-General's Annual Reports. In 1875–6, the telegraphic correspondence of the country resulted in 20,973,535 messages; in 1885–6 it amounted to 39,146,283; and in 1895–6 it produced no less than 78,839,610. Each decade has thus practically seen the number of public telegrams doubled, and this, as regards the later period, in the face of the facilities afforded by telephonic communication between town and town. But whatever may be the demands in this direction, or in relation to railway signalling, in no branch should that viii *Application of Electricity to Railway Working.*

expansion be attended with greater advantage to all who are interested in the advancement of our railways; than in its application to lighting and power.

To the Institution of Civil Engineers and to the Institution of Electrical Engineers, the Author tenders his thanks for the permission accorded him to avail himself of the data contained in certain papers read by him before those Institutions. To Mr. W. H. PREECE, C.B., F.R.S., for the use of tables on the stress of wires, &c.; and to the Editors of the *Electrician, Electrical Review, Engineering*, and the *Engineer*, for extracts from those Journals, his thanks are equally due, and as heartily tendered.

W. L.

DERBY : *October* 20, 1896.

CONTENTS.

CHAPTER I.
ON THE CONSTRUCTION OF A LINE OF TELEGRAPH.

PAGE

Description and character of stores—Poles—Specification for poles —Pole arms—Pole brackets—Line insulation—Terminal insulators—Conductivity—Iron wire—Copper wire—Staying wire —Stay rods—Stay blocks—List of stores 1

CHAPTER II.
SURVEYING.

Bases of line— Number of wires— Height of poles—Distance between poles—Height of wires—Where crossing roadways— Where crossing railway lines—Poles—Arrangement of arms —Staying—Selection of route—Survey book 30

CHAPTER III.
CONSTRUCTION.

Laying out stores—Erection of poles—Staying—Stay spurs—Terminations—Wire guards—Tension of wires—Tables of sags and stresses—Jointing—Wood casing 35

CHAPTER IV.
TELEGRAPH INSTRUMENTS AND BATTERIES.

Single-needle—Bright's bell—Sounder—Duplex—Single and double current keys— Telephones — Phonopore — Induction— Overhearing—Hygienic telephone—Batteries 59

CHAPTER V.
BLOCK SIGNALLING.

Definition—Length of Section—General adoption—Single and three wire systems—Tablet and electric staff—Systems in operation on various lines—The single-needle block on double lines—On single lines—Spagnoletti's—Preece's three-wire—Preece's one-wire—Tyer's—Pryce and Ferreira's—Remarks . . 79

CHAPTER VI.
SINGLE-LINE WORKING.

Observations—Tyer's tablet block—Webb and Thompson's electric staff system 122

CHAPTER VII.
AUTOMATIC BLOCK SIGNALLING.

Considerations on the employment of automatic signalling—Timmis' system in use on the Liverpool Overhead Railway—Electro-pneumatic system—The Hall system—Principles . 143

CHAPTER VIII.
INTERLOCKING.

Observations — Principles — Single-needle block interlocking — Siemens' hydrostatic contact maker—Saxby and Farmer's interlocking system—Sykes' system—Tyer's interlocking block —Tyer's signal lock 155

CHAPTER IX.
MISCELLANEOUS APPLIANCES IN CONNECTION WITH BLOCK SIGNALLING.

Signal repeaters—Repeater discs—Light indicators—Train indicator—Signal instruments for level crossings—Lightning protectors—Movable bridges 190

Contents. xi

CHAPTER X.
ELECTRIC LIGHT AND POWER.

Advantages—Incandescent lighting—Arc lighting—Value in goods yards—Aid to traffic—Economy—Plant on Midland Railway —Selection of site for lamps—In goods yards—In goods sheds —Height of lights—Cost—Electric light and gas compared— Gas plant—Concentration of plant for lighting and power purposes 213

CHAPTER XI.
TRAIN LIGHTING.

General remarks—Trains in operation on Brighton company's line—Experiments on Midland Railway—Principles—Dynamo —Mode of driving—Arrangement of apparatus—Coupling— Arrangement of lights—Auxiliary engine and dynamo—Stone's system 238

CHAPTER XII.
INTERCOMMUNICATION IN TRAINS.

Existing systems—Remarks thereon 258

CHAPTER XIII.
ADMINISTRATION.

ENGINEERING.

General remarks—Chief control—Subordinate officers—Construction staff and work—Maintenance staff and duties—Store depôts—Method of accounts—Examination and testing of wires—Electric light and power—Supervision 260

TRAFFIC BRANCH.

Arrangement of circuits—Reorganisation of circuits—Unnecessary use of wires—Order of precedence for messages—Code time —Collection and examination of messages—Use of check sheets—Train reporting—Telegraph message code . . . 272

xii *Application of Electricity to Railway Working.*

APPENDIX.

	PAGE
USEFUL RULES for the guidance of attendants and others engaged in electric lighting duties	293
PRECAUTIONARY INSTRUCTIONS in relation to the avoidance of, and for the treatment of sufferers by, accidental shock from electric current	295
SPECIFICATION FOR TELEGRAPH POLES	300
SPECIFICATION FOR GALVANISED IRON WIRE	302
SPECIFICATION FOR COPPER WIRE	306
SPECIFICATION FOR STRANDED STAYING WIRE	308
SPECIMEN LEAF OF SURVEY BOOK	309
SPECIMEN LEAF OF TRAIN TABLET EXCHANGE BOOK	310
STORES ACCOUNT FORMS	311–323
INDEX	325

ILLUSTRATIONS.

FULL-PAGE ENGRAVINGS.

LINE OF H POLES, SIX WIRES ON AN ARM . . .	*Frontispiece*
LINE OF TELEGRAPH POLES FITTED WITH IRON ARMS .	*To face page* 8
LINE OF H POLES, FOUR WIRES ON AN ARM . . .	,, 38
TERMINAL POLE WITH LEADING-IN CUPS	,, 42
GIRDER BRIDGE CARRYING SUSPENDED ARC LIGHTS . .	,, 223

ILLUSTRATIONS IN THE TEXT.

FIG.		PAGE
1	TUBULAR IRON ARM	8
2	WOODEN ARM, EARTH-WIRED	10
3	POLE BRACKET	11
4	TERMINAL BRACKET	11
5	INSULATORS	13
6A	TERMINAL INSULATORS, IMPROVED FORM	16
6B	TERMINAL INSULATORS	17
7A	SHACKLE INSULATOR	18
7B	FLETCHER'S TERMINAL SHACKLE	18
7C	FLETCHER'S TERMINAL SHACKLE, IN POSITION . . .	18
8	NO. 11, G. I. BINDER	24
9	BINDER AND TAPE FOR COPPER WIRE	25
10	RATCHET STAY-ROD	27
11	POLE-LIFTER	38
12	H-POLE, BRACED AT FOOT	39
13	H POLE, STAYED	40
14	ATTACHMENT OF STAYS TO POLE	42
15	STAY-SPUR	43
16	LEADING-IN CUP, DOUBLE	43
17	LEADING-IN CUP, SINGLE	43
18	WIRE GUARD	44
19	WIREMAN'S THERMOMETER	45
20	IMPROVED WIRING TONGS	49
21	IMPROVED WIRE BARROW, WORKED BY ONE MAN . .	50
22	IMPROVED WIRE BARROW, WORKED BY TWO MEN . .	50
23	JOINTING CLAMP	52

xiv *Application of Electricity to Railway Working.*

FIG.		PAGE
24	Copper Wire Joint	53
25	Boxing for Covered Wires	54
26	Flush Box	56
27	Iron Tubing, Joint	57
28	Single-Needle Instrument	60
29	Single-Needle Coil	61
30	Revolving Wires	66
31	Revolving Wires, Position on Poles	67
32	Hygienic Telephone, in Section	69
33	Hygienic Telephone, Perspective	69
34	Block Bell Telephone	70
35	Block Bell Telephone Switch Pillar	71
36	Block Bell Telephone Switch Pillar Connections	72
37	Phonopore, Diagram	73
38	Dial of Single-Needle Block	88
39	Single-Needle Block Instrument	89
40	Single-Needle Block Instrument, Interior	89
41	Pegging Arrangement	90
42	Trigger Lock	90
43	Diagram of Block Circuit	91
44	Spagnoletti's Block Instrument	93
45	Screen of Spagnoletti's Block Instrument	94
46	Preece's Semaphore Block Instrument	95
47	Preece's Semaphore Block Instrument Switch	96
48	Preece's Semaphore Block Instrument Indicating Bell	96
49	Preece's Three-Wire Block, Diagram of Connections	97
50	Preece's Single-Wire Block	99
51	Preece's Single-Wire Block, Side Elevation	100
52	Preece's Single-Wire Block, End Elevation	101
53	Preece's Single-Wire Block, Releasing Coil	101
54	Preece's Single-Wire System, Diagram of Connections	102
55	Tyer's Block Instrument	104
56	Tyer's Block Instrument, Three-Wire	106
57	Pryce and Ferreira's Three-Wire Block	107
58	Pryce and Ferreira's Three-Wire Block, Internal Arrangement	108
59	Block Bell	110
60	Block Bell, Interior	111
61	Block Bell Key	111
62	Block Bell, Indicator	112
63	Block Bell, Relay	113
64	Junction Diagram	116
65	Staff for Single Line Working	122
66	Staff Ticket	123
67	Staff Ticket Box	125

Illustrations. xv

FIG.		PAGE
68	Tyer's Tablet Instrument	127
69	Tyer's Tablet Instrument, "No. 5"	133
70	Tablets	134
71	Tablet Pouch	134
72	Switching-Through Apparatus	135
73	Webb and Thompson's Electric Staff Instrument	137
74	Webb and Thompson's Electric Staff Instrument, Interior	139
75	Automatic Block Signalling, Liverpool Overhead Railway	147
76	Electro-Pneumatic System	149
77	Hall System—Diagram	150
78	Hall Automatic Signal	151
79	Hall Automatic Signal, Interior	152
80	Hall Automatic Signal, Electrical Part	153
81	Single-Needle Interlocking Block	159
82	Single-Needle Interlocking Block, Interior, Side Elevation	160
83	Single-Needle Interlocking Block, Locking Part	160
84	Single-Needle Interlocking Block, Signal Lock, Side Elevation	161
85	Single-Needle Interlocking Block, Signal Lock, Plan	162
86	Hydrostatic Rail Contact Maker	164
87	Hydrostatic Rail Contact Maker, Affixed to Rail	164
88	Hydrostatic Rail Contact Maker, Vertical Section	165
88A	Hydrostatic Rail Contact Maker, Mercury Vessel	166
89	Saxby and Farmer's Interlocking Block Instrument	168
89A	Saxby and Farmer's Interlocking Block Instrument, Interior, Side Elevation	168
90	Saxby and Farmer's Interlocking Block Instrument and Frame	169
91	Saxby and Farmer's Commutator	170
92	Saxby and Farmer's Rail Contact Maker	172
93A	Saxby and Farmer's Electric "Slot"	173
93B	Saxby and Farmer's Electric "Slot"	173
93C	Saxby and Farmer's Electric "Slot"	174
94	Sykes' Interlocking Block Instrument	176
95	Sykes' Interlocking Block Instrument, Interior	176
96	Sykes' Interlocking Block Instrument, Interior	177
97	Sykes' Interlocking Block Instrument, Interior	178
98	Sykes' Electric "Slot"	181
99	Sykes' Rail Contact Maker	182
100	Spagnoletti's Interlocking	184
101	Tyer's Interlocking Block	186
102	Tyer's Interlocking Block Commutator	187
103	Tyer's Automatic Signal Lock	186

xvi *Application of Electricity to Railway Working.*

FIG.		PAGE
104	ELECTRIC REPEATER, CONTACT MAKER	192
105	ELECTRIC REPEATER, POSITION OF CONTACT MAKER ON SIGNAL POST	193
106	ELECTRIC REPEATER, DIAGRAM OF CONNECTIONS	193
107	REPEATER INSTRUMENT.	194
108	REPEATER DISC	196
109	REPEATER DISC, RINGING ARRANGEMENT	197
110	EXPANSION BAR	198
111	OIL LAMP EMPLOYED IN SIGNALS, WITH EXPANSION BAR IN POSITION	199
112A	LIGHT INDICATOR INSTRUMENT	200
112B	LIGHT INDICATOR INSTRUMENT, INTERIOR	201
113	LIGHT INDICATOR CONNECTIONS	202
114	TYER'S TRAIN INDICATOR	203
115	TYER'S TRAIN INDICATOR, ELECTRICAL CONNECTIONS	204
116	TYER'S TRAIN INDICATOR, ELECTRICAL CONNECTIONS	204
117	TYER'S TRAIN INDICATOR, ELECTRICAL CONNECTIONS	205
118	LEVEL CROSSING INDICATOR.	206
119	LIGHTNING PROTECTOR FOR TERMINAL INSTRUMENTS	209
120	LIGHTNING PROTECTOR FOR INTERMEDIATE INSTRUMENTS.	209
121	PLATE LIGHTNING ARRESTER	210
122	LIGHTNING ARRESTER, VACUUM TUBE	211
123	ELECTRIC LIGHTING, GIRDER BRIDGE.	220
124	DIAGRAM, ILLUMINATING POWER OF LIGHTS AT VARIOUS HEIGHTS	223
125	DIAGRAM, ILLUMINATING POWER OF LIGHTS AT VARIOUS HEIGHTS	225
126	IRON HOOD FOR ARC LAMPS	228
127	FOUR-SIDED LANTERN FOR ARC LAMPS	229
128	LAMP PILLAR	230
129	PLAN SHOWING POSITION OF ELECTRIC AND GAS LIGHTS IN GOODS SHED	233
130	TRAIN LIGHTING, CIRCUIT ARRANGEMENT.	244
131	TRAIN LIGHTING, GUARD'S VAN	245
132	TRAIN LIGHTING, ELECTRICAL COUPLING	246
133	TRAIN LIGHTING, ELECTRICAL COUPLING COMPLETE.	247
134	TRAIN LIGHTING, ELECTRICAL COUPLING, SECTION	248
135	TRAIN LIGHTING, COUPLING BOX.	249
136	TRAIN LIGHTING, POSITION OF LIGHTS	250
137	TRAIN LIGHTING, POSITION OF LIGHTS	251
138	TRAIN LIGHTING, POSITION OF LIGHTS	252
139	TRAIN LIGHTING, CONNECTIONS	253
140	TRAIN LIGHTING, "BRUSH" DYNAMO.	240
141	POLE DIAGRAM	266
142	CODE-TIME DIAGRAM	279

THE

APPLICATION OF ELECTRICITY TO RAILWAY WORKING.

CHAPTER I.

ON THE CONSTRUCTION OF A LINE OF TELEGRAPH.

IN the construction of a line of telegraph the description and character of the stores to be employed forms an important element. It is not unusual to see, especially on newly established lines of railway, material of the cheapest and, consequently, most meagre description made use of. A telegraph is necessary. The line has, under Board of Trade Regulations, to be worked by the block system, and hence certain wires must be provided. It is, in all probability, a contract matter, the main object being to provide that which will meet the requirements necessary to the passing of the line and no more. As a rule the provision of the telegraph, in common with the equipment of the permanent way and signals, is in the hands of the engineer entrusted with the construction of the line of railway, and unless he has experience in such work, he is only too glad to place the construction of the telegraphs in the hands of a contractor who professes to a knowledge of all that is required. To how large an extent it would be to the advantage of the railway company into whose hands the line is eventually to come, to construct by their own staff, under the supervision of their own electrical engineer, such telegraph lines, can call for but slight demonstration at my hands.

In the first place, it is clear that the capacity of the line will be limited to the accommodation of the number of wires required; that this number of wires will also be reduced to the fewest possible; that the scantling of the timber will, short of its rejection, be as small as the conscience of the contractor will permit. No sooner, however, does the line come into active work than demands for more wires arise, when the short and slim poles have to be replaced by others, with the result that, practically, the cost of rebuilding the line, plus that of the poles displaced, has to be incurred by the incoming company. Although reference is here made to the poles only, it will be understood that a like result will attend such alterations in respect of many other items.

The employment of inferior material, or of that material which, regardless of durability, meets the requirements of the moment at least cost, is a mistake. Timber which readily decays, or that which is less durable for the purpose in view; galvanised iron wire of a small gauge, as that which is known as No. 11, in comparison with that of a larger gauge, as No. 8; even the best galvanised iron wire in preference to a much smaller gauge copper wire, is in each instance an error. A creosoted pole will last much longer than a plain or unpreserved larch pole; a No. 8 gauge galvanised iron wire will last much longer than a No. 11 galvanised iron wire; and a copper wire—copper being in our ordinary atmosphere very durable—will last much longer than an iron wire. We save, by employing the better and more durable material, not merely the cost of the material but the cost of the labour in renewal. Thus, say a mile of poles will cost £20, material and construction. If the poles are of larch or unpreserved wood, they will probably be *hors de combat* in the course of ten years; whereas, if creosoted poles were used they would last thirty or even forty years if not required to be removed for other reasons than that of renewal. In employing the more durable, if even slightly more costly, of the two kinds of poles, we thus save the cost of labour incurred in the renewal of the poles twice over, plus that of the cheaper class of pole—say, at all events, £30 per mile in the thirty years. We must also not lose sight of the fact that the cost of renewal is greater than the cost of the original construction. With renewal we have to clear away the old as well as to install the new. The same law applies to the wires.

The employment of good material combined with good workmanship will always be productive of better results than will that of an inferior article. Let the material selected be of the best, and the work carried out in a thoroughly stable manner.

It is not possible to construct lines of telegraph capable of withstanding all the vicissitudes of an English climate without incurring an unreasonable outlay—probably not even without regard to cost; but it is possible to so construct them that they shall withstand all but the most exceptional of our climatic conditions. When snow gathers upon the wires and becomes converted into ice, the weight alone is frequently sufficient to break the wires, but when this condition is attended by a strong wind, the wires or the poles must succumb. The only means of preventing this is to strengthen the poles and to employ heavier and stronger wires. This would involve a larger increase in expense than could with reason be entertained. It is, perhaps, needless to point out that where a choice exists it is preferable for the wires than for the poles to succumb. With good staying and a judicious provision of the line, the preservation of the poles may generally be effected.

In the following remarks it is not proposed to enter upon a full discussion of all the minor items of material employed in the construction of a line of telegraph, but rather to consider broadly those which are of the greater importance and which mark the latest advances in their application to the subject under review.

Poles.—The description of pole now almost universally employed for telegraph lines is the ordinary red fir of Norway. It should be felled in the late autumn or winter months, when the sap is down, stripped, and allowed to season during the next year. When perfectly dry and well seasoned it should be creosoted. It may then be forthwith made use of, but it is better to allow it to lie for a few weeks in order that the oil may coagulate in the grain of the wood. If the pole is erected too soon after being subjected to the creosoting process, a quantity of the oil will run from the wood, and, in a measure, be wasted.

When a quantity of poles, sufficient to constitute a cargo, is required, it is desirable to arrange with timber shippers for such lengths as are required to be cut in the forests, shipped and delivered in the spring. On delivery from the vessel the poles should not be allowed to remain longer in the water than seven days, less if possible, and as soon as they can be got on to the stacking ground they should be arranged in such a manner as will insure a free circulation of air around them. When stripped of the bark they should again be stacked to still further season and dry.

In specifying for poles it is usual to stipulate that they shall be:—

(1) Of Norway Red Fir, straight, free from knots, and cylindrical in form.

(2) That they shall be cut from the butt end of the tree.

(3) That they shall not be less than a given diameter at the top, nor less than a given measurement at a distance of 6 feet from the butt end.

(4) That they shall be felled when the sap is in the roots.

(5) That they shall not be allowed to lie in water prior to being shipped, nor when being unloaded, more than seven days.

(6) And that each pole shall be cleanly cut at a right angle to the length of the pole across its butt end.

In *branding* and *stamping* poles, either with the initials of the company or the date of felling, or the year in which they have been creosoted, it is a good plan to cause this to be done at a given distance from the butt end of the pole, say ten feet; the engineering officer is then able to see, when the pole is erected, the extent to which it is planted in the ground.

It is important that *poles about to be creosoted* should be dry, and that the requisite quantity of creosote should be injected or pressed into them.* It is asserted that when poles are wet or charged with moisture, on their being placed within the creosoting cylinder and the exhaust applied, the moisture can be extracted from the wood before the creosote is allowed to enter the cylinder. This is very doubtful. Probably a great deal of surface, or near surface, moisture may be so extracted, but where the inner grain of the wood has become to any extent charged, it is very improbable the exhaust which can be applied exercises sufficient influence to affect that part of the structure. The pressure under which the creosote is applied, however, undoubtedly does good, and probably where the degree of moisture is such as to admit of the penetration of the creosote in combining with the dampness in the fibre, it to a great extent counteracts the destructive action of the moisture.

When poles are creosoted in a *wet* condition the moisture held in the grain of the wood is driven in towards the heart of the tree; the outer portion of the stick accepts the creosote and is preserved, but the inner portion remains charged with moisture. In time fermentation is set up, the fibre, saturated with wet, becomes a mass of pulp, and if the pole is struck with a tool of some kind it will be found to emit a hollow sound. Cases have occurred where, due to this condition, the heart or core of the stick has been drawn out of the outer shell, leaving the latter in the form of a hollow tube.

It is not necessary that telegraph poles should be creosoted completely through. It is better that the heart of the tree should not be affected. If the creosote penetrates to the outer layers of the core, it will have formed an effective shield against decay from external causes, and the pole will still

* The quantity of oil of creosote required by the Post Office to be injected into poles varies from 10 to 12 lbs. per cubic foot of timber. The Midland Railway requires 8 to 10 lbs.

retain its natural elasticity. If creosoted entirely throughout every fibre it will, when the creosote has become consolidated and dry, as it will in the course of a few years, become what is termed "short," and more liable to break. If the pole is dry and the creosote injected in proper quantities and under the necessary pressure, the sap wood of the pole is sure to be fully creosoted. If of red wood, i.e. Norway red fir, the core does not need creosoting.

Decay of Poles when planted in the ground, almost invariably arises with that portion buried in the soil. Some highly interesting and important experiments conducted by Mr. W. H. Preece, the present Engineer-in-Chief to the Post Office, fully demonstrated that the destruction of the pole at "the wind and water" line, i.e. the ground line, is due to a continuous capillary action. The moisture at the foot of the pole finds its way up each tiny fibre of the pole—drawn there by the influence of the higher temperature above the soil — and there, expanding under the same influence, bursts asunder its cell and escapes as vapour into the atmosphere, the outer fibres being those first affected. The pole thus in course of time becomes—the sap wood rapidly, the core less rapidly— completely disintegrated and destroyed.

This is what takes place in plain or unprepared timber, such as larch. With creosoted timber moisture is excluded— it cannot penetrate the fibre charged with the oil—and we never see such a pole suffering decay in this manner. We may gather from this how desirable it is that when planting creosoted poles the bottom of the pole should not be cut off, so as to expose any portion of the fibre which may not have been creosoted to the joint influence of the moisture of the soil in which the pole is planted and the warmer atmosphere above ground.

When it was the practice to plant unpreserved timber, and poles rotted, as they did, at the ground line, it was usual and advisable, periodically, to open out the soil around the foot of the pole for the latter to dry, and after having cut away the decayed portion, to serve it over with a coat of tar and pitch ; but with creosoted poles this is quite unnecessary. If we remove the soil from around the buried portion of a creosoted pole, we find the condition of that portion which has been buried in the earth practically the same as when it was planted : the creosote has not been extracted or evaporated from the fibre, as is, perhaps, the case with that portion of the pole exposed to the sun and rain and atmosphere. The creosote is retained in the fibre, and to serve it over with a further preservative is only waste of labour and material. Where the upper portions of such poles become bleached by exposure to climatic influence, it may with advantage be served over with a composition formed of one part Stockholm and three parts coal tar, well boiled.

There are, naturally, certain positions where creosoted timber is, from æsthetic or other reasons, objectionable. Pitch pine is the most durable timber for employment where a square-cut pole is required. It should be as free as possible from dry knots and sap wood. That portion to be planted in the ground should be served over with two coats of preservative compound, composed of Stockholm tar, $\frac{1}{2}$ gall.; pitch, $\frac{3}{4}$ lb.; coal tar, 2 galls.; Russian tallow, $\frac{1}{2}$ lb.—laid on hot. The Post Office employs for the same purpose the following composition : Stockholm tar, $1\frac{1}{2}$ parts ; gas tar, well boiled, 7 parts ; slaked lime, $1\frac{1}{2}$ parts ; with, where desired in order to give the poles a gloss, a little linseed oil. Every portion of the part to be buried should be well coated, especially the butt end cross section.

In thus dwelling upon the treatment of poles and the cause of their decay, the author feels that he has been guilty of a slight digression, but he at the same time believes that the introduction of the subject at this point may be acceptable to the reader, and had it not been introduced here it is probable an opportunity for reference to it elsewhere might have been lacking.

Iron Poles have been employed to a very slight extent for telegraph purposes in England, and still less on railways than on the road lines maintained by the British Postal Telegraph Department. Creosoted timber is even more durable than wrought or rolled iron ; it is also cheaper. Hence the reason.

Still there are instances where some cheaply made iron poles have been employed for light lines. Angle iron of the same section throughout has formed the pole ; the arms, when of wood, being slightly notched to fit that side of the iron bored for the reception of the bolt.

There is no stability in such poles. They are useless without stays. The only reliable iron pole is that formed of tubular iron, and these for railway work can, in respect of cost, in no way compete with creosoted poles.

Pole Arms.—It has been the practice for a number of years to employ only the best English oak for pole arms. The difficulty which, at all events in some parts of the kingdom, attends the production of British oak for the purpose, has raised the cost to such an extent as to induce users to search for something which may take its place.

A large portion of the Post Office main trunk telephone line between London, Glasgow and Edinburgh, where it passes over the Midland railway, has been equipped with *iron arms*. These are the outcome of the difficulty attending the production of English oak arms, and their substitution, in appearance and in respect of durability, leaves nothing to be

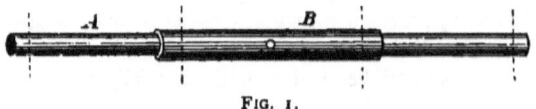

FIG. 1.

desired. Fig. 1 is a representation in perspective of the class of arm thus used, and the plate, which gives their general effect when in position, is from a photograph of the line referred to. It will be observed that the centre portion of the arm is strengthened by a closely fitting tube of larger diameter than the longer tube which forms the complete length of the arm. This outer tube is driven on when heated, the mass is then galvanised and the arm is complete. A 40-inch G.I. tubular arm takes the place of a 44-inch wooden arm, there being no need to extend the iron arm, as is the case with that formed of wood, in order to avoid splitting at

[To face p. 8.

LINE OF POLES FITTED WITH IRON ARMS.

the insulator bolt hole. Arms thus made can, it is clear, be produced any length or size. That represented in the figure is formed of wrought-iron tubing $\frac{1}{8}$ inch thick and $1\frac{3}{4}$ inch in diameter for the longer section A, and of 2-inch tubing $\frac{1}{8}$ inch thick for the larger section B.

Iron arms, formed of L iron, have also been employed experimentally. They are less costly than the tubular arm, but fall short of the degree of strength which the latter enjoy. The 40-inch tubular arm referred to above has been tested up to a strain of 600 lbs., applied at the outer insulator bolt hole, i.e. 1 inch from its extremity. On attaining this stress both the arm and the bolt which secured it to the pole became bent.

The advantages of the tubular arm are: (*a*) greater durability; (*b*) greater strength; (*c*) economy in space in packing; (*d*) greater security for workmen when standing on them when fixed; (*e*) less cutting away of poles for their reception; (*f*) no earth wiring.

The disadvantage, if it is a disadvantage, rests in the fact that being metal it forms a ready conductor, and any wire which may become detached from its insulator will, the arm being connected to *earth*, give practically *dead earth*. Whether this is not an advantage rather than a disadvantage is a question. A partial earth always presents some difficulty in localisation, a dead earth is more readily traced; still a circuit may be worked through a partial earth, if not of too pronounced a character, whereas when to full earth it cannot be. Experience will show us which is best in this respect. It is perhaps worthy of notice that although the line referred to has been in use now some time, no inconvenience has attended the employment of the iron arms.

Where the arms are of wood they should be cut from the heart of the tree, free from sap, shakes and knots. Large numbers of arms are cut from Stettin or American oak, which, although not so durable or so reliable as English oak, is exceedingly serviceable. Whether of English or foreign wood, they should be well seasoned before being either painted or earth wired. The tannic acid of English oak, if not evaporated, is an active agent in the destruction of the earth wires where such are used.

In some quarters, and, it is believed, in the British Postal Telegraph Service, the *earth wiring* of arms has been abandoned. The object of this earth wiring is to prevent leakage from any one wire to any other wire on the same arm. Recent improvements in the insulating properties of insulators have, no doubt, removed to a great extent the danger of such leakage; still the possibility, where no earth wiring exists, especially in wet weather and with dirty insulators, has not been wholly removed. It is also probable these earth wires fulfil a part in conducting atmospheric electrical discharges to earth.

Fig. 2 illustrates a method of earth wiring which has been found very efficient for the purpose in view. It provides for full connection with the arm bolt, which is in the usual

FIG. 2.

manner connected to the earth wire at the back of the pole, while the portion of the arm earth wired is practically covered by the pole, so that any wire which may from any cause become loose cannot form contact with it.

The arms may be of any length desired, cut, chamfered and bored to meet the requirements of the service demanded. Usually, for ordinary construction, 24 inches by 2½ inches by 2½ inches, and 33 inches by 2½ inches by 2½ inches are the lengths employed for poles with wires arranged two on an arm. For terminal arms the horizontal thickness is usually increased by half an inch, and the length of the arm limited to 24 inches. We thus have for a terminal arm the dimensions 24 inches by 3 inches by 2½ inches.

For telephone lines it is desirable the wires should be run in a *square*, each wire a given distance apart one from the other, and so run that they shall revolve around one another. This renders the employment of four wires on an arm convenient; the four wires forming a square can then be arranged one foot distant one from the other, the arms being fixed on the pole one foot apart, centre to centre; and the insulator bolt holes being also a foot apart. It is quite possible that in the near future, especially as copper wire becomes more generally used, the provision of four wires on an arm, whether for telegraphs or telephones, will become very general. It is, therefore, desirable in determining the description of arm to be used, to consider the propriety of such an arrangement. For four wires on an arm, with single poles, a 44-inch arm will be found serviceable. For H poles the arms should be 56 inches, and for six wires 72 inches. If iron arms are used these lengths may be reduced by 4 inches.

Instead of arms, *pole brackets* (Fig. 3) are at times used. The pole arm is preferable. Where brackets are employed they should be tubular and galvanised. Three-inch clout nails or screws are necessary to fix them to the poles. They should be arranged alternately on either side of the pole.

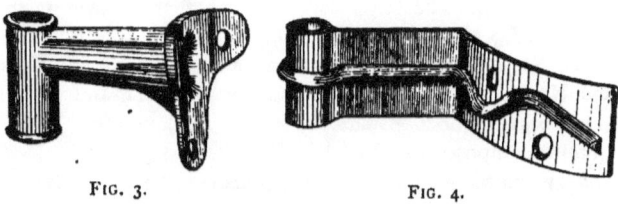

FIG. 3.   FIG. 4.

For terminations a form of *malleable iron bracket* (Fig. 4) is frequently very suitable. Wooden terminal arms fixed longitudinally, i.e. in the direction of the wire, are equally good where the pole is of such scantling as to admit of the wires being sufficiently far apart to avoid contacts in rough weather.

Insulators.—The efficiency of an aërial insulator is the resistance which it affords to the passage of the current between the point at which the line wire is attached and the bolt, or support, of the insulator. The *metal*, or ware, of which it is formed should be effectually vitrified; it should be impervious to moisture, and possess a smooth, well-glazed surface.

The best insulators of the present day are formed of highly glazed porcelain. Earthenware insulators are still used to a large extent by some railways. They are less costly than porcelain, but it is doubtful if, even viewed from this standpoint, their employment is attended with any degree of economy; for where they are used on long circuits the defective insulation during wet and damp weather, especially during the winter months, largely reduces the working capacity of the wire, while on short (*block*) circuits, worked by constant current, the leakage must result in the employment of more battery power than would otherwise be necessary.

A mixed class of insulators on a pole is an eyesore, but where insulators of two colours, as white and brown, are used, the effect is still more objectionable.

The descriptions of insulator chiefly employed are, Fig. 5, the Post Office form (*a*); the corrugated (*b*); and the earthenware class (*c*). The latter consist of what are known as the No. 8 D.V.; No. 11, D.V.; and the Z. The No. 11 is similar to the No. 8 in character, but of a smaller type. The No. 8 and No. 11 are employed respectively for No. 8 and No. 11 gauge wire. The Z insulator has been employed for "block" circuits composed of No. 8 or No. 11 gauge wire. Most railway companies have now relinquished the use of No. 11 galvanised iron wire as generally unremunerative, and with it, of course, the No. 11 insulator.

As already remarked, the best insulator is that which affords the highest resistance between the point at which the wire carrying the current is attached and the iron bolt of the insulator. Mr. John Gavey, of the British Postal Telegraph Service, in a paper read before the Institution of Electrical Engineers, has laid down the following principles for securing this result:—(1) By increasing the length to be traversed by the current; (2) diminishing the section of the conducting film; (3) retention of a dry surface on one portion of the insulator; (4) a form which will not aid or retain deposits of dust, soot, &c., nor foster spiders and other spinning insects.

A consideration of these important points will explain the

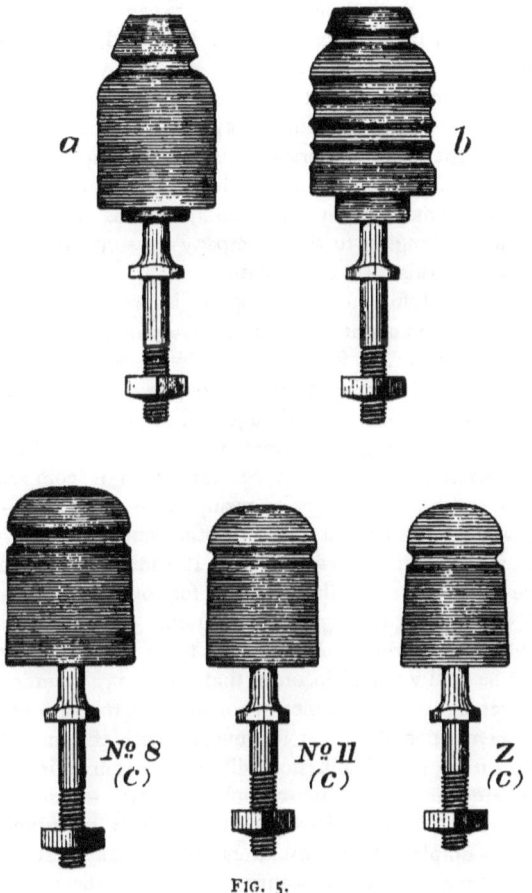

FIG. 5.

principle pursued in the construction of the corrugated form of insulator. The rounded shoulder of the insulator, with its plane surface at that point, as also the plane portion below the corrugations, are retained as parts which, in addition to the edges and exposed portions of the corrugations, may be washed by the rain ; the corrugations perform a double duty, viz. that of extending the distance between the crown of the insulator and its junction with the bolt, and in providing a finely rounded edge on which it is practically impossible for dirt to accumulate, a further object of which is to dissipate readily any moisture which may have gathered upon it. Some of the most important wires belonging to the Post Office, working between Edinburgh and Glasgow and London, where they pass over the Midland system, as also all important wires belonging to that company's system, are insulated by means of this form of insulator.

The Post Office form (a, Fig. 5) is practically the same divested of the advantages of the corrugations.

The earthenware form of insulator, known as the D.V. (double Varley) is composed of two parts, an inner and an outer shed, or cup, cemented together. The best description of cement for fixing the several parts, as also the insulator bolt, is what is termed *black* cement, which is composed of 5 parts smith's ashes, 4 parts resin, and 8 parts fresh-water sand washed clean of all dust and other impurities, the whole well cooked over a slow furnace. It should be well mixed before being placed in the cauldron for cooking, and must be kept well stirred during that operation. It is important it should not be burnt. All the pieces to be cemented together should be well warmed in order that there may be no chilling of the cement where it comes in contact with the various parts. If the cement is chilled at its juncture with the insulator or the bolt its power of adhesion will be greatly impaired.

Where cement of a hygroscopic character—such as Portland cement, plaster of Paris, or others which are mixed with water—is employed, the insulators will in many instances in cold weather, when the moisture stored up in the cement becomes frozen, burst, and thus produce leakage of current during damp weather. The power to resist this disruption on the part of the insulator is entirely a question of the power of the ware to resist the force exercised by the crystallisation of the moisture held by the hygroscopic cement.

The system of fixing the bolts in the insulators by cement, as also that of forming the insulator of more than one shed or cup, is gradually dying out. All modern insulators are made in one piece and provided with bolts which screw into the insulator, as represented by a in Fig. 6A.

The employment of a screw bolt aids insulation. Of two sets of insulators in other respects equal, the one fitted with fixed (cemented) bolts, the other with screw bolts, the latter will be found to possess the highest insulating power.

Insulators should always be tested for insulation before being issued. The cups should be washed and wiped clean before placing them in the vats prepared for this purpose. These vats are shallow cases or boxes, lined with lead, filled with slightly acidulated water to within half an inch of the edge of the insulator cup. The cups, after being placed within these vats, are filled with the acidulated liquid, allowed to remain from 10 to 12 hours, and are then tested by means of a high battery power—not less than 100 cells, better by 200 cells—and a delicate reflecting or other galvanometer, for leakage between the insulator bolt and the vat.

If any deflection is observed on the galvanometer the cup should be removed from the vat, rewashed, examined, and if no flaw is apparent, reinserted in the vat and again tested. If the deflection is reproduced the cup should be rejected—although not apparent to the eye of the observer, it is defective, due either to imperfect vitrification or to a flaw. There is usually one part of the insulator which is not glazed. This part should of course be inserted in the liquid. Where this is not so a small portion of the glaze should be filed away, so as to expose the ware itself to the influence of the liquid. If not properly vitrified it will absorb moisture, and the testing officer will readily obtain a deflection on his galvanometer. Usually it will be found that the crown or top surface of the insulator cup, as also some portions of the bolt hole, are left unglazed.

The insulator is as important a factor in the successful

working of a line as is the conductivity of the wire. It should be carefully treated. Any dirt which may adhere to the inside of it when erected will remain there. It cannot be removed by the rain as dirt which may adhere to the outside. All insulators should be wiped clean before being placed in position on the arms. Any which have been set together by means of a hygroscopic class of cement, when laid out for construction work should be carefully placed on the ground crown upwards, so that moisture may not accumulate within the cup of the insulator and penetrate the cement.

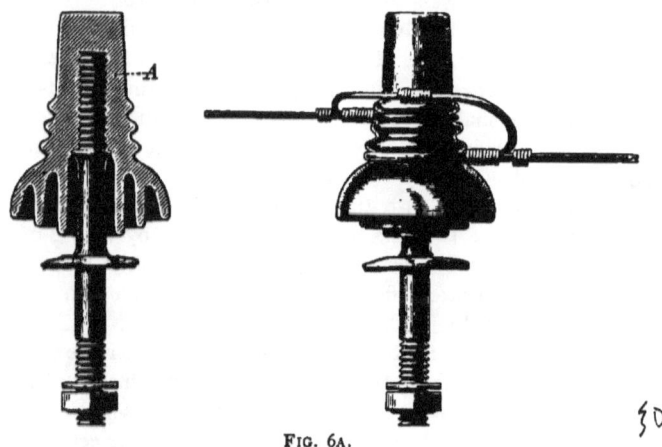

FIG. 6A.

Although a *Terminal Insulator* should at all times prove an insulator, an equally important qualification is strength. It must be capable of resisting the strain thrown upon it by the termination of the wire, and its construction and arrangement should be such that, should the porcelain ware become broken by stone-throwing or other causes, the wire shall not lose its hold, but shall be retained in its position sufficiently to prevent it from interfering with other wires attached to the same pole. Figs. 6A and 6B represent the latest form of terminal insulators. The former, known under the author's name, is the most recent. It will be observed that each is fitted with a bolt, much heavier and stronger than that employed in the ordinary line insulator, and that it is provided with an extended and equally strong shoulder, the purpose of which is to aid in sustaining it in its vertical position when fixed in the terminal arm. The point at which the wire is, in the later form A, attached is, as will be seen from the illustration, brought nearer to the base, thereby reducing the leverage strain. Grooves are also provided for the accommodation of varying gauges or descriptions of wire: thus, if it were desired to renew a section of No. 11 G.I. wire by No. 8, if one of these insulators were placed in the position of the

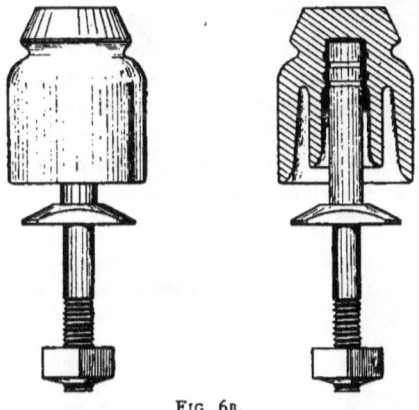

FIG. 6B.

ordinary line insulator and the No. 8 wire terminated on the lower groove and the No. 11 on the upper groove, the tensile influence upon the insulator bolt would be fairly equalised, and it should, the spans being about the same, retain its upright position. Again, assuming connection has to be made between an iron and a copper wire, the one, say the iron, would be terminated on the lower groove and the copper wire on the upper groove, the two being joined by a loop. This form of terminal is made with a screw bolt or cemented bolt: the former affords the best insulation, and the experience so far gained suggests that it is equally as strong as that with the cemented bolt. The advantage of the screw bolt A over the cemented bolt is that the only safe cement by which the bolt can be fixed being the *black cement* previously alluded to, one of the ingredients of which, resin, is liable under the influence of high summer temperatures to soften, the insulator cup may be drawn out of the vertical, whereas with the screw bolt fitting the thread of the insulator closely, there is no such possibility. The deviation of the insulator cup from the vertical would not materially impair its property as an insulator, but it would be objectionable in appearance.

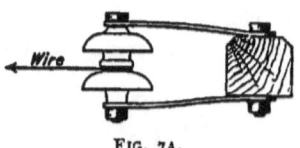

FIG. 7A.

The shackle cone, Fig. 7A, is largely used by railway companies for terminations. It is an indifferent insulator. The wire is necessarily attached in the centre, and leakage, when the insulator is damp or dirty, takes place between that point and the strap both top and bottom.

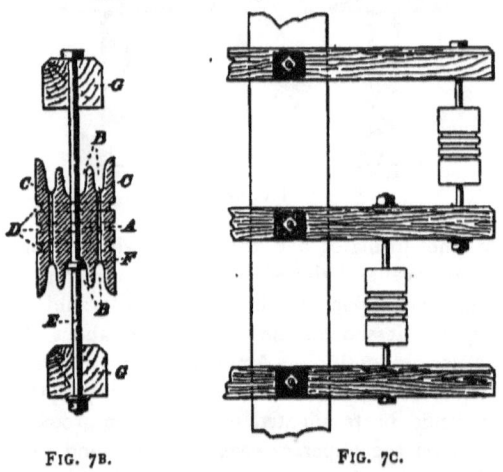

FIG. 7B. FIG. 7C.

An improved form of shackle is that invented by Mr. G. Fletcher, of the London and North Western Railway, represented in Fig. 7B, where A indicates the porcelain ware of which it is formed; B, channels for increasing the surface distance between the point of attachment for the wire and the shackle bolt; C, holes for discharging any accumulation of moisture which may collect in the channels; D, grooves for attachment of line wire; E, shackle bolt, extending from the arm G above to the arm G below the shackle. The bolt E is, as is shown in the figure, provided with a shoulder F, on which the insulator rides. The mode of application to the pole is shown in Fig. 7C.

Wire.—Although copper wire was employed for the earliest telegraph erected, viz. that between Slough and London, it was rapidly superseded by iron, the latter being more economical and, at the time, sufficiently suitable to the purpose. Galvanised iron wire followed, and has for many years been universally adopted. Strange as it may appear, the author believes it to be the fact that as the original iron rails laid down by some of the earlier railway companies proved more durable than those of later date, so the iron wire first employed for telegraph purposes has proved far more durable than even the galvanised wire of the present day. In 1871, Mr. A. Graves, of the North Eastern Railway, experimentally erected some copper wire in the Newcastle station yard, where, owing to the smoke, the iron wire rapidly deteriorated. The main effort in the introduction of copper wire, however, dates to the erection of a copper wire by Mr. W. H. Preece, between London and Newcastle, in 1884-5. The erection of this wire afforded practical proof that the employment of copper wire for telegraphic purposes not only reduced resistance but afforded increased capacity, and reduced that electro-magnetic inertia which had proved so inimical to fast speed telegraphy. These efforts were influential in directing attention to the desirability of the more general employment of copper wire for telegraph purposes. Large quantities of copper wire have within the last fifteen years been erected on the Midland and a few other lines of railway, attended in every case with satisfactory results. A line of heavy copper wires, ranging in gauge from No. 4 to

No. 8, has also recently been erected over the Midland company's system between London and Settle (Yorkshire), as part of the Post Office Trunk Telephone system, between London and the North, Belfast, &c. Two 400-lb. copper wires are now being erected by the same company, at the instance of the Postmaster-General, between London and Carlisle, for, it is understood, telegraph purposes.

Iron Wire.—For many years past the gauges of iron wire employed have been No. 11, No. 8, and sometimes, for very special wires, No. 6 or No. 4. The use of No. 11 is now practically abandoned. In the open country, free from smoke, it may prove economical, but these conditions are not obtainable apart from lines of railway, and certainly they do not exist on the route of the locomotive. An iron wire reduced by corrosion, or whatever may be the cause, to No. 12 gauge should be renewed, therefore its life is the period between the date of its erection and its reduction to this size. This will vary with local conditions. The cost, in labour, for its renewal will be the same as that for the renewal of No. 8. If we allow No. 8 to be reduced to the same gauge, we shall find that, broadly, a No. 8 wire will last as long as two No. 11 wires. We thus save, by the employment of No. 8, instead of No. 11, the labour cost of one renewal—an economy which, on a large mileage, is worth consideration. There are, of course, other items which might be brought into account, but the saving indicated in the labour alone is sufficient to justify the abandonment of the lesser gauge.

Copper Wire.—At the same time there would appear reason to regard the days of galvanised iron wire as numbered. Iron wire, unless protected by some covering impervious to moisture, will rust, and in doing so will degenerate in *conductivity* as well as in *tensile strength.* Copper is far more durable. When first erected in an ordinary atmosphere an oxide forms upon its surface, but otherwise there would seem to be little change. So far it is not possible to speak with any degree of certainty of the life of copper wire, but it would appear at least reasonable to conclude that it will be much more durable than iron. In first cost copper is more expensive. We have therefore to discount it to that extent. Hitherto No. 14 gauge—100 lbs.—has generally been employed in place of iron wire. It is questionable if No. 14 is a sufficiently large gauge. No. 12½ (150 lbs. to the mile) is probably that which for ordinary railway telegraph service will be employed. Its cost the reader will readily be able to compare with that of iron. Its advantages are : higher conductivity, less electro-magnetic inertia, longer life, reduced weight and strain upon poles. In tensile strength it is inferior to new No. 8, but there is a period in the life of the latter when its tensile power is, by corrosion, reduced below that of the 150-lb. copper. Thus, in the long run, it is possible that the copper wire is in tensile power as good as the No. 8 G.I. wire. It would also appear from such observations as have been made by those interested in the question, that although snow gathers as rapidly upon copper as upon iron wire, its conversion into ice—which is usually the cause of the breakage of wires when snow adheres to them—takes place to a less extent with copper than with iron wire.

As previously observed, there is in existence on the Midland lines a large mileage of copper much of which has been in use for several years—the earliest of that now standing having been erected in 1879. Snowstorms have been encountered by these wires, and they have withstood the test of one of the most severe winters experienced in England. A great deal is of No. 14 gauge, none is of a larger gauge than 12½. They have certainly stood as well, if not better, than the iron wires, among which they are, in many places, interspersed.

There are important factors which contribute towards this result. A copper wire of No. 14 or No. 12½ gauge is, of course, much smaller than a No. 8 gauge wire. The mass of metal is less. Its power of retaining heat is, consequently, in point of mass, less. Copper is a more ready conductor of heat, or cold, than iron. When the atmospheric condition is such as to encourage falling snow to cling to the wires, the circumstances are as follows :—The wire must be slightly warmer than the snow. If it were below freezing temperature the snow would not adhere to it. The heat stored up in the wire melts the snow which first falls upon it, when, the wire becoming chilled, the melted snow is formed into ice. According to the degree of warmth in the wire, and its power of parting with it, the snow is melted, and falls from, or clings to it in a half frozen condition. If there is much warmth in the wire the snow at first becomes melted and falls from it; but if the snow is persistent, a moment arises when the temperature of the wire is chilled to such an extent as to convert the snow into ice. We can readily conceive how the *rapidity* of this chilling process governs the formation of the ice— whether forming of it a mere shell, or, as is at times the case, a solid mass of ice from an inch to two inches in diameter.

Thus we see how the extent of surface, the mass, and the specific heat capacity of the two descriptions of wire each play their part in this transition of snow into ice. There is every reason to believe that copper wire will not gather ice upon its surface to the same extent as will iron wire—gauge for gauge—but where the copper wire is of a smaller gauge than the iron wire, the advantage should be largely in favour of the former. All points considered, there is no doubt that copper possesses many advantages over bare galvanised iron wire.

The author has, from time to time, employed, to a large extent, a form of covered galvanised iron wire which is well worthy of mention. The wire is first covered by layers of cotton and then drawn through a bath of what is known as *West's Composition*, so as to thoroughly saturate the covering. It is necessary it should be erected when this covering is in a moist condition, i.e. before it solidifies; but if protected from the air it will remain moist for a considerable time. The West's Composition is a chemical compound, the ingredients of which are only known to the manufacturers, Messrs. John Fuller and Son. Its peculiarity, in comparison with other covered wire in the coating of which oily compounds are employed, is that whereas in the latter the virtue of the covering is gradually extracted and vaporised by the atmo-

to Railway Working. 23

sphere and warmth of the sun, the West's Composition simply hardens and becomes impervious to moisture, and this condition it retains for a very long period.

There are some important points to bear in mind in dealing with wire covered in this manner. It should not be used too quickly after being covered, but allowed to stand for a week or so, to allow the surplus liquid to drain off. Care should be taken in its erection that the covering is not stripped. It is a good plan, when running it off the wire barrow, to allow it to pass between the folds of a piece of leather grasped in the hand of the man in charge of the barrow. This gives it a compressed form, robbing it of any rough edges or excrescences, which, if allowed to remain, would cause the wire when erected to look rough and unworkmanlike. It should not be drawn over the pole arms, but run over sheaves or rollers attached to the arms. It is not well to draw it out over the ground or grass, as it will strip the covering. Care must be observed in pulling it up to the proper strain. If pulled up too tight, when the covering has become hardened and cold weather supervenes, the wire will be stretched, and this may cause the covering at some point to open. Wherever the covering opens moisture will penetrate and rust will ensue. This wire is bound in with No. 16, covered in a similar manner, and where jointed the joints have to be lapped by hand with cotton yarn, which is supplied already saturated for the purpose. Not only is wire covered with this preparation very useful for smoky localities, but in certain chemical districts, where the air is charged with corrosive fumes, it is practically invaluable.

Binding Wire.—Although the author has for many years abandoned No. 16 wire for binding the line wire to the insulator, employing in its place a special description of No. 11, there are still many who make use of it. Where employed it should be of a soft and flexible character, and should not "scale," i.e. the galvanising should not flake off when twisted around No. 11 wire. The wire will, under any circumstances, in a very short time oxidise, and when this is the case the oxide will travel down the sides of the insulator, and as oxide of iron is, when moist, a conductor, it will largely impoverish the insulating properties of the insulator. If the wire "scales," the oxidation will arise almost immediately it is used.

The best binder for galvanised iron wire is a No. 11 G.I. wire, specially made so as to render it free from scaling when laid around a wire of its own gauge, and of a soft and ductile nature, so as to admit of its being readily laid around the wire which it has to bind to the insulator. Five laps around the line wire on either side of the insulator C, as seen at B, Fig. 8, will be found sufficient to hold the line wire, in case of its severance, from running back more than three or four poles, and that only to a slight extent. Its advantages are: less cost in material and labour; less chafing of the line wire; improved insulation; greater power to resist gales of wind and snow; longer life; and greater security.

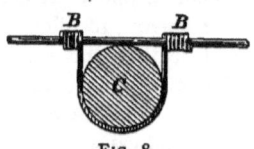

FIG. 8.

Binding Wire for Stays.—Where stays are bound by means of wire other than that composing the stay itself, a stiffer class of No. 11 G.I. wire than that advocated for binding the line wire to the insulator should be used. If it is too ductile the strain thrown upon it by the stay may cause it to yield, when the binding would become useless. Ordinary conductivity No. 11 is quite suitable, and if, as should in all cases be done, it is served over with a coating of Stockholm tar and grease at the same time as the thread of the ratchet stay rod is served, it will not rust.

Binders for Copper Wire.—The Post Office have recently introduced binders for copper wires, which follow the practice pursued by the author for so many years past in relation to binders for iron wires. There is this difference in dealing with the binding-in of copper wires. It has been found advisable to lap the line wire with binding wire at the point where it rests against the insulator, in order to protect it from chafing. This is quite independent of the binding itself. The Post Office no longer use binding wire for this purpose, but a specially made tape or copper strip C, Fig. 9.

·This is laid around the line wire A first, and the binder D, which consists of a piece of copper wire rolled flat at either end, varying in gauge according to the size of the line wire,

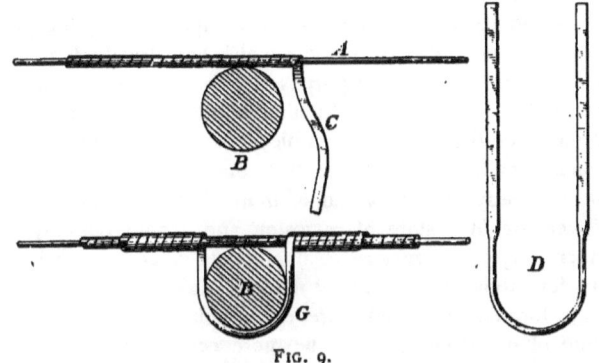

Fig. 9.

is then laid around the insulator B, and the flattened ends lapped around the wire already overlaid by the copper tape. Of the flattened ends of the binder, one end takes an under

COPPER TAPE BINDERS.

Weight of Wire per Mile.		Length of Binder.	Copper Tape.	
Line.	Binder.		Length.	Width.
Lbs.	Lbs.	Inches.	Inches.	Inch.
800	400	19	24	¾
600	400	19	24	¾
400	200	19	23	3/16
200	150	17	22	4/32
150	150	17	22	3/32
100	100	17	22	3/32

lap and the other an over lap, so that when being drawn tight by the wireman's pliers, the tape which has been laid around the line wire shall be drawn tighter rather than loosened.

In *binding wire for jointing* there has been of late years no change. No. 16 G.I. wire is still used for all Britannia joints in G.I. wire ; and No. 17 tinned copper wire for copper wire joints.

Staying Wire.—With some companies it is the practice to use seconds wire, i.e. wire which has already done duty as line wire, for staying purposes. No greater mistake can be made ; the strength of a line depends mainly upon the strength of the stay. To employ seconds wire is to employ a hand-made stay, and one the constituent parts of which have already been deprived of more than half their tensile power, are in a state of corrosion, and cannot possibly last much longer. Every renewal of stays means so much outlay for labour, and if a stay made of new wire will last two stays formed of seconds wire, even though the seconds wire were of no value, it would be more economical to employ that made of new wire.

To attempt to form by hand a stay which is composed of more than one wire is waste labour. The most experienced workman, exercising the greatest care and provided with the most suitable tools, cannot apply each wire so that it shall take its proper share of the strain. Again, conductivity wire is unsuitable for staying purposes. Conductivity wire is capable of a certain amount of stretch or elongation under strain. This is not a desirable condition for a stay : it is required to withstand the strain thrown upon it without stretching and without breaking ; any stretching means a slack stay till again tightened by the lineman. Any degree of slackness jeopardises the stability of the line. The secret of a secure line of telegraph rests in keeping everything as taut as is permissible. The line wires must have a sufficient margin to meet the cold winter weather. There are periods when they will unavoidably hang slack, but the stays should always be taut. When poles begin to rock under stress of wind and the influence of the swaying wires, a weak spot in some one wire will be found, followed by others, ending in a breakdown, which probably would not have occurred had the staying been sufficiently rigid.

Stay wire should be *stranded by machinery*; it should be of a homogeneous or steely character, capable of resisting with very little stretch a comparatively high strain (see Specification in Appendix). Three, five and seven strand will be found the most serviceable. It should always be composed of No. 8 wire, and the *lay* of the twist should not be less than twelve inches. If it has a shorter lay it will be stubborn to handle.

The material thus far dealt with disposes of the chief items required in the construction of a line of telegraph. There still remain, however, numerous smaller parts which scarcely call for special reference here, unless it is the *stay rod* and *stay block*. Of the former the best is the ratchet stay rod, which has been in use now for many years. Its only drawback is the exposure to the influence of the atmosphere of that portion of the threaded rod which, for the time, is not protected by the nut. That portion is exposed to rust, and the thread in time becomes more or less eaten away. To meet this to some extent the author has introduced, and employed with advantage for some time past, a square thread; this does not rust so rapidly as the ∧-shaped thread generally used. The threaded portion of all stay rods should be served over with a composition of tar and grease from time to time.

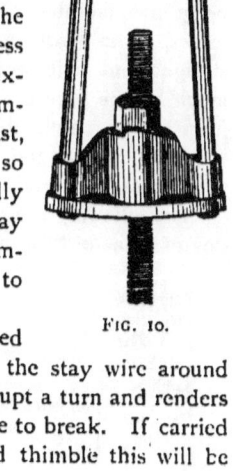

FIG. 10.

It is a good plan to use pear-shaped *thimbles* with all stay rods. To pass the stay wire around the eye of the rod only gives it too abrupt a turn and renders the wires of which it is composed liable to break. If carried around the groove of the pear-shaped thimble this will be avoided. The loop and ratchet of the stay rod is represented in Fig. 10. The ordinary rod is of ⅝ inch section, and that used for terminal poles 1 inch.

Where the thread of a stay rod has become impaired by rust, its holding power may be increased by employing a second nut.

Stay Blocks.—The stay block for all important lines should have at least one flat side. That side should be exposed to the stress, i.e. should be uppermost when placed in the hole at the end of the stay rod. All stay blocks should be creosoted. Second-hand sleepers—sleepers which have done duty but which are still sound—form excellent stay blocks. A 9-feet sleeper will cut three stay blocks, say each 36 inches by 10 inches by 5 inches. They should, after being cut into the required size, be bored for the reception of the rod and then re-creosoted. If the hole for the rod is not bored before the block is again creosoted decay will set in at that point, endangering the future stability of the block. Creosoted blocks buried in the ground are practically indestructible, and, if of the dimensions indicated, will afford a surface resistance of 350 to 360 square inches.

Staples for fixing the stays to the poles should be properly manufactured of No. 8 wire, and not made up of odd pieces, as is frequently the case. The former will drive straight and hold firm; the latter will be controlled very much by the nature and grain of the wood, and will at best form but an unreliable job.

Arm Bolts and other ironwork should be tested for "shortness" of metal. It should bend and not break.

The following is a list of stores required in the construction of a line of telegraph :—

Poles.	Nails, 1½-inch clout, for fixing roofs.
Pole roofs.	Earth wire (No. 8) G.I.
Pole arms.	Staples for ditto (No. 11).
Arm bolts.	,, stays (No. 4).
Pole brackets, tubular.	Stay wire.
Saddles.	,, spurs.
Nails, 3-inch clout { for fixing saddles. ,, brackets. ,, stay blocks. }	Nails for ditto. Stay blocks. 3-inch clout nails for ditto.

Stay rods.	Binding wire for jointing.
Insulators, line.	No. 14 copper or covered G.I. wire
,, terminal.	for crossings.
,, leading-in.	G.P. wire for leading-in.
Screws, 1½ G.I. for ditto.	Wood casing for ditto.
Conductivity G.I. or copper wire.	Screws for ditto.
Binding wire, or tapes and binders for insulators.	Solder.
	Soldering fluid.
Ditto for stays.	Salammoniac lump.

If pole brackets are employed, arms and arm bolts will not be needed. If saddles are not used, the pole roof will be fixed by 1½-inch nails, otherwise the 3-inch clout nails employed for fixing the saddle will serve for fixing the roof. Several of the items may not of course be needed; the object of the statement is to afford a list of such articles as may be required, from which a selection of what is necessary for the special work in hand may be made.

CHAPTER II.

SURVEYING.

BEFORE proceeding to make a survey for a line of telegraph, it is necessary to determine the basis upon which the line is to be constructed—the number of wires it shall *ultimately* accommodate ; the distance between the poles on the straight part of the line and around curves ; the height which the lowest wire, when the poles have their full complement, shall be from the ground, and the height above public roadways over which the line may pass ; the arrangement of the pole arms, whether the wires shall be placed two or four, or in what manner, on the arm ; the distance the arms shall be apart one from the other, centre to centre ; at what point the stay wires shall be attached, and to what extent the line shall be stayed.

An important point is also the headway required for railway vehicles where wires cross the main lines. Although the overbridges are not nearly so high, yet in some cases it is considered necessary that wires which cross lines of railway should afford a clear headway of 18 feet from the rail level. This is considered desirable in order that a man standing on an engine tender may not be brought into contact with the wire. It is contended, and with much reason, that whereas a man in such a position can see sufficiently in advance of him a bridge which the train may be approaching, he cannot see an isolated, or possibly a number of wires in sufficient time to get out of their way.

Ordinarily, poles on lines of railway are planted 80 yards apart on the straight, and 75 yards on the curved portions of the line. For lines carrying wires of heavier gauge than

No. 8, the distance between the poles should be correspondingly lessened—say 60 yards on the average, or 30 poles to the mile.

Where important lines are carried over public roadways it is desirable to place a pole on either side of the public road so as to prevent the wires crossing the roadway, in case of breakdown on either side, slacking down so as to impede vehicular or other traffic passing underneath. It is also a good plan to solder the wires to the binders at such points, so as to avoid any chance of their being drawn through the binders.

Where a large number of wires have to be erected it is desirable to consider the propriety of constructing an H line, i.e. two poles braced together somewhat as represented by the letter from which the character of the line derives its title. The distance at which such poles are planted apart the one from the other should not exceed that of the ordinary single line of poles, but unless other objections intervene, more nearly approximate to a distance of 60 yards.

The employment of copper wire will, as a rule, call for no departure from these provisions. Experience has shown that No. $12\frac{1}{2}$, and even No. 14 hard drawn conductivity copper wire will stand very well spans of the extreme length named.

Poles are usually armed—the top arm 8 inches from the top of the pole,* the remaining arms 12 inches apart, centre to centre of arm bolt. The arrangement of the arms upon the poles is perhaps to some extent a matter of taste. The author has preferred to employ a 33-inch arm for the top arm, followed by a 24-inch, and so on throughout the equipment.

It is here worthy of consideration whether it will not, as has been previously indicated, be found advisable to erect wires four on an arm. Many of the Post Office lines are now so arranged, the purpose being the accommodation of telephone wires. With the object of overcoming the effect of induction between neighbouring wires, it is desirable to erect them so that each one of any four wires shall revolve the one

---

* Where poles are armed for four wires on an arm the upper arm should be 9 inches from the top of the pole.

around the other. Thus, where the insulators are arranged 1 foot apart, the arms also being 1 foot apart centre to centre, we get a square formed by the four wires constituting two metallic circuits, and by a systematic course of erection these wires may be made to revolve in the same manner as the strands of a rope. The distance between wire and wire, and, the gauge being the same, the inductive influence will be equal between wire and wire, and as the wires by the manner in which they are caused to revolve, cross one another, so will the inductive influence of the one part be counteracted by that of the other part. This can only be effected conveniently by arranging the wires two on each side of the pole, or where H poles are employed, in devoting the four inner wires to the purpose ; or by employing arms long enough to accommodate six wires, two on each of the outer sides of the poles and two between the poles.

The extent to which poles should be stayed, and the strength of the stay to be employed, must be contingent upon the number of wires to be erected, the position of the line, if exposed or sheltered, and other local conditions. Poles carrying 5 to 10 wires should be stayed both sides on the straight, and against the curve on the curved sections of line. When the number of wires exceeds ten the poles should be stayed both sides, on the curve as well as on the straight. In making this provision the future requirements of the line should of course be borne in mind, as also the natural degeneration of the material, due to rust, &c.

The basis upon which the line is to be constructed having been determined, it remains to carry out the survey, and under it to decide the positions in which the poles shall be placed ; the length of pole required ; number of arms, bolts, &c. ; whether to be stayed or not, and if on one side or both ; together with other points which may present themselves as the survey is proceeded with, all of which details should be recorded by the surveying officer in the survey book, a specimen of which will be found in the Appendix.

In carrying out this survey one of the first questions which will present itself is, on which side of the line of railway shall the line of telegraph be erected. If time and opportunity will permit, it is well to go through the line in the first instance in order to observe if any, and what, obstructions or difficulties present themselves. Space for staying is an important consideration. The leeward side of the line is preferable to the windward, i.e. the side from which the prevailing winds come. It is, if the poles should ever be blown down, preferable for them to be blown away from the railway metals rather than on to them. The side on which the signal boxes, or the points at which the wires have to be led in for connection with the various kinds of instruments, is a reason for selecting that side. If the line is a single line and there is a probability at a near future date of its being formed into a double line, it is clear that the proper course is to construct the line of telegraph on the single line or completed side, and so avoid having to rebuild the line when the widening takes place. The poles and wires should of course be accessible, therefore they should not be erected in out-of-the-way places. If the position of the signal boxes is indicated, care should be observed not to obstruct the view of the line or signals, as observed from the signal box, or to impair the view of the signals by an approaching train. Poles planted near to signal boxes for leading-in wires should be placed at the back, or at a back corner of the box; they will not then affect the view of the signalman when standing at his signal frame.

It will add to the satisfactory appearance of the line if so constructed that it shall, as nearly as possible, follow the contour of the railway metals, but it must not be forgotten that stability is the first consideration. Harmony of construction is desirable, but security is a more important factor.

The position which each pole should occupy, besides being indicated in the survey book, should also be marked by a stout survey peg being driven into the selected point; and an excellent plan is also to mark the length of the pole required on the fence rail at the spot. A mark of this kind is not only readily seen by the laying out party but is also a guide as to the length of pole required to be dropped off the truck at that point, and in addition tends to check the survey book. The surveying officer has judged the height of the pole required in each instance by his surveying rods, and if any departure is made from the positions in which they have been placed it will lead to confusion, and produce irregularity in the height of the poles when planted. With a well-constructed line on a straight piece of road the poles should stand in perfect line, each one the same distance from the near railway metal ; on a curve it should follow the curve of the line as nearly as may be. Where tall poles have to be employed for crossing roadways, or carrying wires across the line of railway, the poles on either side should lead up to it, so as to graduate the rise and fall. This is necessary, not merely for appearance but for the satisfactory support of the wires.

Where wires have to cross a line of railway or a public roadway it is preferable they should cross at right angles to the road. The span over the roadway should be as short as possible, and if the crossing is over a line of railway, copper (No. 14) wire will be found more durable than iron.

CHAPTER III.

CONSTRUCTION.

WHEN building a line on a newly formed railway it is not always possible to lay the stores out for a complete section of the line by a "special," at least not for such a section as would call for a special train. There are, however, occasions when the line is sufficiently advanced to admit of this course; and there are other occasions, for instance, in the construction of additional lines of telegraph or in the execution of heavy renewals, when a special train is of much value. In loading up such a train it will be attended with convenience if the stores are all loaded in the order in which they are, according to the survey book, required for use. Thus the poles should be loaded in the different lengths as they are, according to the survey book, required to be erected; but if there are sufficient to admit of any one, or even two lengths being loaded on one set of timber trucks, it is better to do so than to mix them. When so many lengths are loaded on one set of timber trucks it is inconvenient for the men to single out and drop off the trucks the shorter lengths.

The other stores, arms, bolts, insulators, wire, &c., should be loaded each into its own truck, a low-sided one; and if the insulators have to be laid out with each pole, they should be grouped together so as to be quickly handed out. It is, however, a good plan to pack these in small casks, a given number in each, and to lay a cask out at a convenient point for the section it is to serve. The insulators can be easily carried out by hand later on as required.

The relative position of each truck load of material depends much upon the line of railway—if in operation or if not yet opened. Where there is plenty of space, and where the small stores can be laid out on one side and the poles and heavy stores dropped off on the other side, it is a good plan to arrange the poles at the rear of the train. When the line is in service, and all the stores have to be laid out on one side of the train, it is desirable the heavy material should be dropped off the trucks first, otherwise it may fall upon the smaller material and injure it. Under such a condition the following is a very useful disposition of the different material.

Locomotive.	Arms.
Empty low-sided goods wagon.	Bolts, &c.
Poles.	Insulators.
Stay blocks.	Wire.
Stay rods.	Guard's Brake.

An experienced foreman will at times lay out poles as the train proceeds at, say, a walking pace. This course of procedure, however, necessitates the poles being loaded one way, i.e. with their butt ends to the tail of the train; and further, the loading of a limited number of poles only on each set of trucks. The bolster-pin at the tail end of the truck, on that side on which the poles have to be laid out, is then cast off, and the man in charge of the truck cases off the required pole as it is called for by the officer in charge of the laying out. The butt end falls to the ground, and as the train slowly proceeds the truck is drawn from under the other end of the pole. Such a system of laying out poles requires great care and experience on the part of the officer in charge.

It is very desirable that all timber trucks loaded with poles for laying out should be "matched," fore and aft, by another timber truck, to form a platform on which the men dealing with the poles may stand; and that the long coupling links which usually connect the trucks together should, for the laying out work, be replaced by shorter couplings, so as to avoid violent jerks to the several trucks forming the train, on each one of which it is probable a man is stationed, and who, if care is not observed, may be thrown out of the truck. These short couplings are of course unsuitable for ordinary service. They should be brought into use only when the

to Railway Working. 37

train is started on its laying out expedition, and great care must be observed when travelling around curves.

A couple of men should be appointed to follow the train in order to gather together and place alongside each pole, the fittings applicable to it which have been dropped from the train; and to see that all heavy stores, such as poles, stay rods and blocks, coils of wire, &c., are placed out of the way of passing trains. Where stores are laid out in the winter months, if the small stores are not laid alongside the poles they are liable to be buried beneath any snow which may fall in the interval between their being laid out and their being required for use.

The train should travel on that side of the line on which the work has to be dealt with, so that, ordinarily, all material may be laid out on the off side and not in the "six-foot."

CONSTRUCTION OF LINE.

It is not the intention of the author to enter upon a detailed description of the various matter involved in the building of a line of telegraph. For such instruction he would refer the reader to Preece and Sivewright's 'Telegraphy,' sixth and later editions, in which the various details will be found to have been fully dealt with. It is true that the subject is there treated mainly from a road line point of view; still in the matter of hole digging, the erection of poles, planting of stay blocks, staying, wiring, &c., the method pursued is, with few exceptions, equally applicable to work which has to be done on railways. To exceptions only attention will be drawn.

Railways differ from roads in that they are more free from abrupt angles, and are formed in cuttings and on banks as well as on the level. The curves are less acute, but, as a rule, the pole setting and staying, together with much other work, has to be carried out on the slopes bordering the lines of railway. Men accustomed to the work find no difficulty in dealing with it under these conditions. "Earth-borers" are unsuitable for other than flat, or nearly flat surfaces, and consequently find no place in railway telegraph work. A short-handled pick with a fairly straight head, supplemented by a strong shovel, are the best tools for hole-digging, except where stiff clay prevails, when a *grafting tool* is necessary. An important point is that the holes, whether for stay blocks or poles, are not dug too large. The less earth removed the quicker will the hole be opened out, the less will be the cost, and the more secure will be the pole, &c., when placed in it.

An excellent tool for use in rearing poles is the *pole lifter*, Fig. 11. It is armed at its extreme end with a double spike, which, as the pole is reared, is thrust into its upper portion. With one pole lifter on either side, and careful manipulation, two men keeping the pole midway between them can readily raise a moderately sized pole into its required position. Ladders are frequently used by the workmen for the same purpose, much to the prejudice of the ladder. A pair of pole lifters are easier to move about with, do their work better, and are less costly.

In erecting H or double poles, it has been the practice of the author to arrange the two poles parallel, and not as is usually done, to give them a greater spread at the bottom than at the top. His object is to enable arms of one length, similarly bored for the reception of the insulators, to be employed throughout. When poles which are not parallel become crowded with wires, either the upper arms have to be much longer than is necessary or in appearance desirable, or the lower arms must be longer than the upper ones.

FIG. 11. On road lines, and on railways where the space at liberty for staying the poles is limited, it is necessary to tie the butt ends of H poles by cross timbers, Fig. 12, not merely in the formation of the structure, but also, by means of the lower cross piece, to obtain a greater hold of the ground and add to the stability of the pole when planted ; but where staying can be applied these tie pieces below the soil may be

[To face p. 38.

TF615
L3

dispensed with without in any way jeopardising the safety of the line. There is in this no great saving in labour, for whereas the H-pole structure calls for a large hole, the staying calls for hole digging for the stays ; but there can be no question that an H-pole stayed on either side is much stronger, so long as the stays are good, than an H-pole which is not stayed, but which is tied by cross timbers under the surface of the soil.

The system is of great advantage where a single line of poles has to be converted into an H-line, Fig. 13. The standing pole should be fairly matched by that which is to combine with it to form the double structure. The additional pole should be planted in an independent hole the required distance from the standing pole, well punned and then allowed to settle. When this is assured it can be slotted for the reception of the arms to match those on the standing pole, and as the arm is fitted to the notch in the new pole, the short arm from the old pole may be removed and the long arm for the double pole take its place. When the poles forming the single line have been uniformly armed, each arm being distant from its neighbour, say 12 inches centre to centre, and the newly planted pole is allowed to become well settled before being armed, a line so constructed looks well and is undoubtedly durable. It will add to the uniform appearance of the line if the arms are fitted to the same side of the poles—say the London side or the north side—throughout.

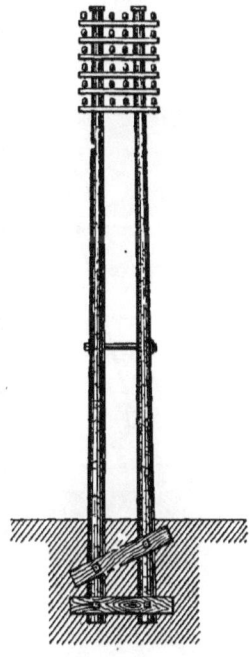

FIG. 12.

It is the practice on some lines to give poles what is termed a *set*, i.e. to allow them to incline a little from the upright against the strain thrown upon them by any curvature of the line. Very little, if any, advantage is gained by this, and it is destructive of the uniform appearance of the line. A line of telegraph, especially on the banks of a railway, is not unsightly if the poles are well in line and regular in height.

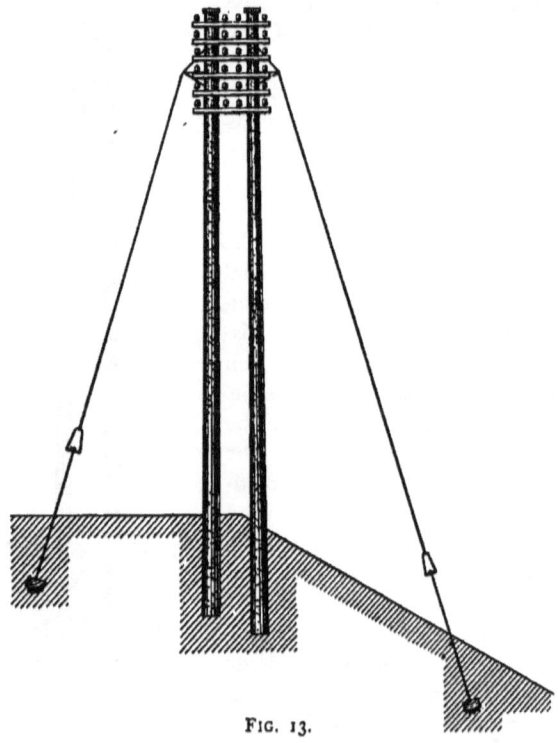

FIG. 13.

but where one pole inclines in one direction and another in the opposite direction, all harmony is destroyed, and any artistic effect which the line might otherwise possess is lost. All poles should be planted upright, and any strain to which they may be subject should be met either by the hold which the pole has in the ground or by the stay.

Poles should be stayed before the wires are erected, otherwise they may be pulled out of upright.

The point at which the stay should be attached to a pole is not always easy to determine. Strictly it should be the *resultant point* of the strain to which the pole may be subjected ; but until the pole has its full complement of wires this cannot be decided. It is therefore better to fix a point which will be serviceable for the future as well as for the present. Between the third and fourth arm from the top of the pole will be found the most serviceable. Where the pole is heavily wired it is advisable to apply a second stay between the eighth and ninth arm, or thereabout, against strains arising from the curvature of the line.

It is an excellent plan to stay heavily wired poles laterally, that is in the direction of the wires, at, say every half mile or every tenth pole. The object of this is, in case of breakage of wires from storm, to meet the lateral strain thrown upon the poles. Where, from accumulation of snow and storm, wires become broken, the sudden relaxation of the tension existing at the moment of breakage, combined with the strain of the wires, throws an enormous stress upon the poles on either side of the break. This strain is transmitted from one pole to another until spent, either breaking the poles or pulling them out of the vertical. The lateral staying of each tenth pole should intercept this stress and save the poles from further damage. Where wires cross roads or railways, the pole on either side should be stayed laterally, i.e. against the road.

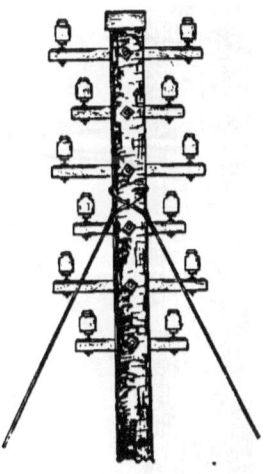

FIG. 14.

In fixing stays to the poles they should be made to cross one another, as shown in Fig. 14 ; and they should, if the space at disposal will admit, have an equal spread, i.e. enter the ground at the same distance from the pole on either side. The greater the spread of the stay the greater service will it render the pole. Properly manufactured staples of No. 4 gauge, not staples made by the gangmen, of No. 8 wire, should be used for fastening the stay wire to the pole. The well-manufactured staple enters the pole in the direction in which it is required, and forms a firm hold. The extemporised staple, having no point, will possibly pursue a really useless direction in the wood and prove of little service.

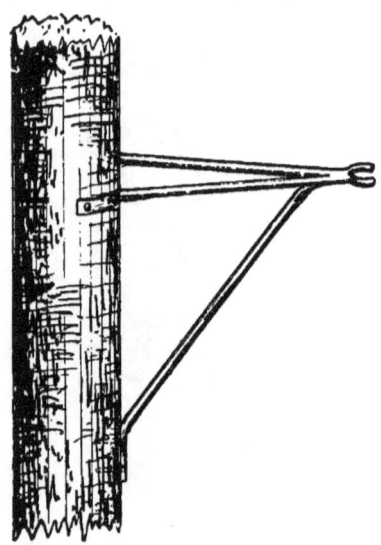

FIG. 15.

Stay spurs, Fig. 15, should be used on H-lines, or where poles are armed with arms of the same length. Under these circumstances the wires are not so far apart as where a short and long arm is used alternately. It is still more desirable to make use of them for poles carrying four wires on an arm.

Where stays are carried over the railway fence they should not be left unprotected: cattle running against them may sustain injury. Each such stay should be accompanied by a piece of timber—a portion of a pole, say 8 feet long—planted 3 to 4 feet in the ground, along with and parallel to the stay rod, which should be bound to it.

Terminal poles should be provided with *eye bolts* for the reception of the staying wires. It must not be forgotten that the terminal pole has to bear the full strain of the wires terminated upon it, and the staying must be corre-

TERMINAL POLE WITH LEADING-IN CUPS.

[*To face p.* 42.

spondingly strong and rigid. If the stay is merely fixed to the pole by being carried around it, as in staying the ordinary line poles, the least movement at the point of attachment will allow the pole to pass out of its upright position. The eye bolt does not admit of any such movement. Terminal poles carrying twenty wires should be provided with three stays, each of seven-strand No. 8 wire.

Where the pole is stout enough it may be grooved between the arms for the reception of the leading-in wires, the groove being neatly covered by a face board; but the usual course is to fix a fillet of wood on either side of the pole inside the arms to carry the *face board*, or cover, and to make use of the recess thus formed for the leading-in wires.

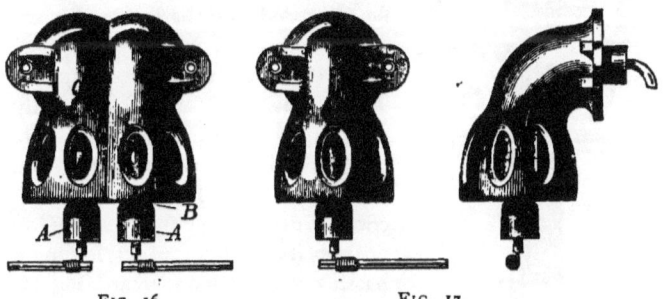

FIG. 16.   FIG. 17.

Leading-in cups, as shown in the figure, protect the gutta-percha wires, and insure better insulation than can possibly be obtained with the use of shackles.

In connecting the copper leading-in wire to the line wire, the latter should, after passing around the terminal insulator, be looped back so as to form a ready attachment for the leading-in wire, which should be lapped around the iron wire some twenty times; the last three or four convolutions only, however, should be soldered. This will provide for disconnection and reconnection, if at any time required for testing or for alterations without lengthening the leading-in wire.

Where the pole is an intermediate leading-in, or terminal pole, double leading-in cups, Fig. 16, may be employed. For single terminations the single leading-in cup shown in Fig. 17 is preferable. The insulating tubes A are narrowed at the point lettered B, for the purpose of keeping the leading-in wire central and avoiding leakage at the edge of the tube.

Where the wires are erected four on an arm, double terminal poles are at times used, or the four wires may be terminated on extra strong terminal arms—arms the lower side or bottom of which have been sheathed with iron—fitted to the front and back of the pole and braced together, the terminal insulators being fixed in the rear arm.

Wires erected on the "revolving" system will, as a rule, need no *guards*, the tendency being in most instances to draw the wire towards the pole. But wires run on the "parallel" should be "guarded" on the inside of every curve, and at all points where they cross over roadways or footpaths. The ordinary "C" guard is usually sufficient for the purpose, but where the curve is a very short one, and there is a possibility, on a wire becoming disengaged from its binding or from its insulator, of its fouling a passing train, a "loop" guard, Fig. 18, formed of No. 8 or No. 4 G.I. wire, fixed around the pole and having within the loop the wire required to be protected, should be used. Care should always be observed in fixing guards that they may not twist round, or otherwise come into contact with the line wire, and so produce a partial earth.

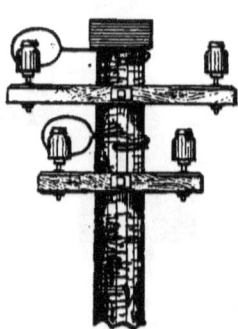

FIG. 18.

Excellent advice on wiring will be found in Preece and Sivewright's 'Telegraphy.' Only recently has the erection of wires been carried out on principles calculated to preserve the wire and at the same time ensure its proper regulation. For years it has been the practice—a practice still pursued in many quarters—for the wireman in erecting wires on a new line to pull them up to a certain tension, practically as tight as he can get them, and then to slack them out to the extent of a couple of notches of the vice. A moment's reflection will show the absurdity of this. In winter the wire is contracted by the cold; in summer it is expanded by the heat. On a summer's day we, between sunrise and sunset, experience a wide range of temperature. The old-fashioned wireman is practically oblivious to this. His rule, whatever it is, is a rule of thumb. It is clear that if we are desirous—and of this there can be no question—that our wires shall be erected in such a manner that they shall not be subjected to a greater strain than they can bear, that we must be guided in the degree of tension to which we subject them by the temperature prevailing at the moment at which they are "pulled up" and regulated. Each wiring gang should be provided with a thermometer, and a wiring card indicating the tension to which the wire should be subjected at the temperature then existing. Fig. 19 represents the form of thermometer which has been employed by the author for years. The glass tube, it will be seen, is protected by a metal back, which is returned on either side so as to guard it from injury by anything being blown against it. The thermometer should be suspended from a wire, or some projection which will enable it to hang in the open, free from any substance which may influence the mercury either by the heat stored up within it or by screening it from the prevailing winds, &c. In point of fact, what we have to aim at is to place it

FIG. 19.

46 *The Application of Electricity*

in a position as nearly as possible analogous to that of the wire which is being dealt with.

The following tables of sags and stresses, which have been issued by the Post Office Telegraph Department, are here inserted by the kind permission of Mr. W. H. Preece.

TABLES FOR STRESSES TO BE OBSERVED IN ERECTING IRON WIRES AT VARIOUS TEMPERATURES.

Temperature in Degrees Fahr.	Stresses in Pounds for Various Spans in Yards.									
	200 lbs. per Mile. (121 mils diameter, No. 10 S.W.G.)					400 lbs. per Mile. (171 mils diameter, No. 7½ S.W.G.)				
	50 yds.	60 yds.	70 yds.	80 yds.	90 yds.	50 yds.	60 yds.	70 yds.	80 yds.	90 yds.
	lbs.	lbs.	lbs.	lbs.	lbs.	lbs.	lbs.	lbs.	lbs.	lbs.
22	135	135	135	135	135	270	270	270	270	270
25	120	124	127	128	130	239	247	253	256	259
30	103	110	115	119	122	205	219	230	238	243
35	91	100	107	112	115	182	199	213	223	230
40	83	92	99	105	110	165	184	198	210	219
45	77	86	94	99	105	153	172	187	199	210
50	72	81	89	95	101	143	161	177	190	201
55	67	77	85	91	97	134	153	169	182	194
60	64	73	81	88	94	127	146	162	175	187
65	61	70	78	85	91	121	139	155	169	181
70	58	67	75	82	88	116	134	149	163	175
75	56	64	72	79	85	111	129	144	158	170
80	54	62	70	77	83	107	124	140	153	165
85	52	60	68	75	81	103	120	135	149	161
90	50	58	66	73	79	100	117	132	145	157
95	48	56	64	71	77	97	113	128	142	154
100	47	55	63	69	75	94	110	125	138	150
Poles per mile	} 35	29	25	22	19½	35	29	25	22	19½

Factor of safety of 4 at 22° Fahr. for all wires.

The stress varies with both the gauge and the material.

The stress for 100 lbs. (No. 14) copper wire is half that for 200 lbs., and the stress for 800 lbs. (No. 4½) copper wire is double that for 400 lbs.

to Railway Working. 47

TABLES FOR STRESSES TO BE OBSERVED IN ERECTING COPPER WIRES AT VARIOUS TEMPERATURES.

Stresses in Pounds for Various Spans in Yards.

Temperature in Degrees Fahr.	150 lbs. per Mile. (97 mils Diameter, No. 12 S.W.G.)					200 lbs. per Mile. (112 mils Diameter, No. 11½ S.W.G.)					400 lbs. per Mile. (158 mils Diameter, No. 8 S.W.G.)					600 lbs. per Mile. (193 mils Diameter, No. 6 S.W.G.)					Temperature in Degrees Fahr.
	50 yds.	60 yds.	70 yds.	80 yds.	90 yds.	50 yds.	60 yds.	70 yds.	80 yds.	90 yds.	50 yds.	60 yds.	70 yds.	80 yds.	90 yds.	50 yds.	60 yds.	70 yds.	80 yds.	90 yds.	
	lbs.	lbs.	lbs.	lbs.	lbs.	lbs.	lbs.	lbs.	lbs.	lbs.	lbs.	lbs.	lbs.	lbs.	lbs.	lbs.	lbs.	lbs.	lbs.	lbs.	
22	120	120	120	120	120	160	160	160	160	160	320	320	320	320	320	480	480	480	480	480	22
25	96	102	106	109	111	129	136	142	145	148	258	272	284	290	296	386	409	425	436	444	25
30	76	84	91	96	99	102	113	121	128	133	204	226	242	256	266	306	338	364	383	399	30
35	65	74	81	86	91	87	98	108	115	122	174	196	216	230	244	262	295	323	346	365	35
40	58	66	73	79	84	77	88	96	106	113	154	176	192	212	226	232	265	294	318	338	40
45	52	61	68	74	79	70	81	90	98	105	140	162	180	196	210	211	243	271	296	317	45
50	48	56	63	69	74	65	74	84	92	100	130	148	168	184	200	194	223	253	277	299	50
55	45	53	59	65	71	60	70	79	87	94	120	140	158	174	188	182	211	238	262	284	55
60	42	50	56	62	67	57	66	75	83	90	114	132	150	166	180	171	200	226	250	270	60
65	40	47	53	59	65	54	63	71	79	86	108	126	142	158	172	162	189	214	238	260	65
70	38	45	51	57	62	51	60	68	76	83	102	120	136	152	166	153	181	205	229	249	70
75	37	43	49	55	60	49	58	66	73	80	98	116	132	146	160	147	173	197	220	240	75
80	35	41	47	53	58	47	55	63	71	77	94	110	126	142	154	141	166	190	212	232	80
85	34	40	46	51	56	45	53	61	68	75	90	106	122	136	150	136	160	183	205	225	85
90	33	39	44	49	54	44	52	59	66	73	88	104	118	132	146	131	155	177	199	218	90
95	32	37	43	48	53	42	50	57	64	71	84	100	114	128	142	127	150	172	191	212	95
100	31	36	42	47	52	41	48	56	62	69	82	96	112	124	138	123	146	167	188	207	100
Poles per mile	35	29	25	22	19¼	35	29	25	22	19¼	35	29	25	22	19¼	35	29	25	22	19¼	Poles per mile

SAGS FOR COPPER AND IRON WIRES AT VARIOUS TEMPERATURES, WHICH PROVIDE FOR A FACTOR OF SAFETY OF 4 AT 22° FAHR.

Temperature in Degrees Fahr.	Sags for Various Spans.																			
	Copper Wire.										Iron Wire.									
	50 yds.		60 yds.		70 yds.		80 yds.		90 yds.		50 yds.		60 yds.		70 yds.		80 yds.		90 yds.	
	ft.	in.	ft.	in.	ft.	in.	ft.	in.	ft.	in.	ft.	in.	ft.	in.	ft.	in.	ft.	in.	ft.	in.
22	0	8	0	11⅜	1	3⅝	1	8⅜	2	1⅞	0	9½	1	1⅝	1	6½	2	0¾	2	6⅝
25	0	9⅞	1	2¾	1	5¾	1	10½	2	4	0	10¾	1	2⅞	1	7⅞	2	1½	2	8
30	1	0½	1	4¼	1	8⅜	2	1⅝	2	7⅛	1	0½	1	4¾	1	9¾	2	3½	2	10
35	1	2¾	1	6⅝	1	11¼	2	4⅜	2	10	1	2	1	6½	1	11⅝	2	5⅜	2	11⅞
40	1	4½	1	8¾	2	1⅝	2	6⅞	3	0⅝	1	3½	1	8	2	1¼	2	7⅛	3	1¾
45	1	6¼	1	10¾	2	3¾	2	9¼	3	3¼	1	4¾	1	9⅜	2	2⅞	2	8⅞	3	3½
50	1	7¾	2	0¾	2	5¾	2	11⅜	3	5½	1	6	1	10⅞	2	4¼	2	10⅜	3	5⅛
55	1	9⅛	2	2¼	2	7⅝	3	1⅜	3	7¾	1	7	2	0⅛	2	5¾	2	11⅞	3	6¾
60	1	10½	2	3⅝	2	9¼	3	3¼	3	10	1	8⅛	2	1¼	2	7	3	1⅜	3	8¼
65	1	11¾	2	5¼	2	11	3	5¼	3	11¾	1	9⅛	2	2½	2	8⅜	3	2¾	3	9¾
70	2	0⅞	2	6⅝	3	0½	3	7	4	1¾	1	10⅛	2	3½	2	9½	3	4⅛	3	11¼
75	2	2	2	8	3	2⅛	3	8⅝	4	3⅝	1	11	2	4⅝	2	10¾	3	5⅜	4	0⅝
80	2	3⅛	2	9¼	3	3½	3	10¼	4	5⅜	1	11⅞	2	5⅝	2	11⅞	3	6⅝	4	2
85	2	4¼	2	10⅜	3	5	3	11⅞	4	7⅛	2	0¾	2	6⅝	3	1	3	7⅞	4	3¼
90	2	5¼	2	11⅝	3	6¼	4	1⅜	4	8¾	2	1⅝	2	7⅝	3	2⅛	3	9⅛	4	4⅝
95	2	6¼	3	0¾	3	7⅝	4	3⅜	4	10⅜	2	2⅜	2	8⅝	3	3⅛	3	10¼	4	5⅞
100	2	7⅛	3	1⅞	3	9	4	4⅜	5	0	2	3⅛	2	9⅜	3	4⅛	3	11⅜	4	7⅛
Poles per mile	35		29		25		22		19½		35		29		25		22		19½	

It will be observed that the variations in sags and stresses under changes of temperature are much greater in short spans than in long ones, the stresses between the low winter and high summer temperature, with a copper wire, being reduced in the ratio of about 2 to 1 with a 90-yard span, while the reduction is nearly 3 to 1 in the case of a 50-yard span. One thing to be learnt from this is that to avoid ultimate excessive winter strains, much more care is necessary in the erection of short than in that of long spans.

When wires are strained up too tightly—i.e. when sufficient *sag* is not allowed—the advent of cold weather will stretch or break them. Good iron wire will stretch, at various points and to a considerable extent. We may therefore have in each span a point where it has by this action been attenuated, or reduced from a No. 8 gauge to a No. 11 gauge. This means that at a future day the wire will break at that point. Copper wire has much less stretching capacity, and will therefore, in all probability, break on being subjected to the increased stress. Another important point in relation to galvanised iron wire, worthy of consideration, is whether any stretching to which the wire may be subjected has not an injurious effect upon the galvanising; whether it does not impoverish its value by opening it out, and thus inaugurate the destruction of the wire.

Coils of iron wire when laid out upon a line of railway should not be allowed to rest upon cinder ashes. All wire, whether iron or copper, should be paid out from a wire barrow.

An excellent description of *tongs*, for use in place of the ordinary draw vice, has recently been brought out by Mr. Harradine. Its great advantage is that it is adaptable to any gauge of wire, and that its grip of the wire extends throughout the section of the grip pieces. It does not, as with other tongs or vice, grip the wire at one point only : where this is the case, naturally the wire becomes indented, bruised or otherwise damaged, which not unfrequently leads to breakage at that point.

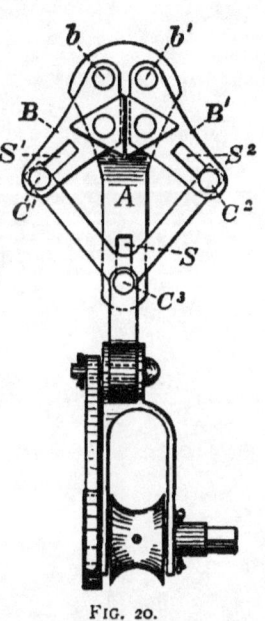

FIG. 20.

The tool coupled to the ratchet portion of an ordinary draw vice is shown in Fig. 20. A is the base plate on which the other parts work ; B, B^1 are wing pieces slotted at S^1, S^2 ; C is an angular piece, rigid, but free to move in the slot S ; studs C^1, C^2 couple C to the wings B, B^1, and these wings are attached to the base plate by the studs b, b^1, but are free to

E

move thereon. At C³ the rigid angular piece is controlled by a flanged stud, so that its movement may be confined to

FIG. 21.

the slot S, and it is at the same spot attached to the hook, or loop, which couples the tool to the draw vice. Fixed on B, B¹ are two triangular pieces, free to revolve on the flanged studs which confine them to the wings B, B¹. These

FIG. 22.

triangular pieces are slightly grooved and milled to three different sizes to accommodate various sizes of wire.

E 2

If now the tool be held in the hand, and the base plate A moved towards the draw vice, the wing pieces B, B¹ will open at C¹, C² and carry with them the triangular pieces which form the jaws of the vice. The wire to be strained up is now placed in that groove most suitable to it, the wing pieces are pressed together, and the wire is at once gripped throughout the length of the sides of the triangular pieces between which it has been placed. The wire is thus held without injury, and the ratchet portion can be employed in the usual manner for drawing up the wire to its required degree of tension.

The most recently constructed wire barrow is shown in Figs. 21, 22. The drum on which the coil of wire is placed

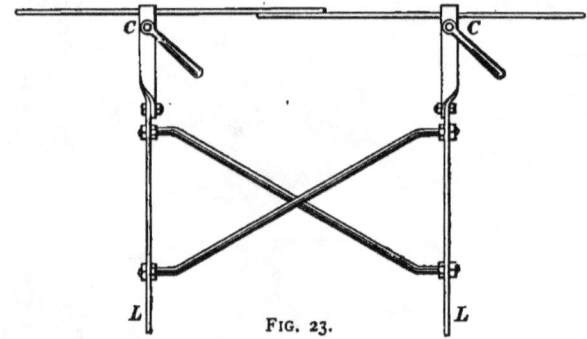

FIG. 23.

is arranged to travel vertically, not horizontally, as in the old form; the wire thus falls direct on to the ground over which the barrow travels. A brake regulates the speed at which the drum revolves. It is made with a wheel similar to a garden barrow, but is also equipped with a pair of portable handles, so that if required to be carried fore and aft, that course may be pursued.

In front of the drum a pair of clamping pieces are arranged for jointing. This jointing frame can be used as shown on the barrow, or its legs L may be placed in the earth or on a bank, and the joint made there. The ends of the wires to be joined are first cut off square, well cleaned, and then placed

to Railway Working. 53

each within its clamp C, Fig. 23, so that they shall overlap each other as follows: No. 14, 1¾ inch; No. 12, 2 inches; No. 11, 2¼ inches; No. 8, 2½ inches; No. 6, 2¾ inches; No. 4, 3 inches. The clamps are then tightened up by means of the screws seen at C C, and the joint is ready to be bound. It will be observed that this makes no provision for the usual bent up end of the line wire. The need for this no longer exists. It is possible that originally its object was to assist the solder, that too much stress should not be imposed upon it. Possibly the solder now used is of a harder character. Under any circumstances the bent ends are no longer needed. Where defective soldering occurred the bent up ends rendered the tracing of the defect difficult. It is better a defective joint should at once draw asunder, than remain a source of annoyance and difficulty in working.

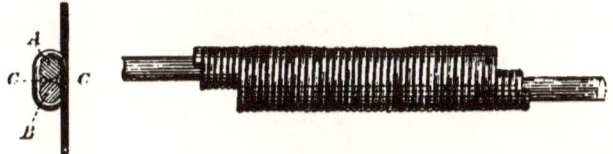

Fig. 24.

It is important the joints of copper wires should not be overheated. It is therefore necessary the soldering iron should not remain in contact with the wire longer than is absolutely necessary to effect a proper liquefaction of the solder; and the less time occupied in this the better. Towards this end the Post Office, in dealing with the larger gauge wires, viz. No. 4 and No. 6, pack the space between the two wires A B by a piece of smaller wire inserted on either side, as shown in Fig. 24 at C. The pieces of packing are cut the length of the joint, and after the first few laps of the binding wire have been laid on are pushed in under the binding. The binding wire taken in the middle of its length is laid around the wires first in one direction, then in the other, finishing the joint as shown in the figure. A suitable flux, such as *Baker's soldering fluid*, is then applied, and a well-faced soldering iron, which must be sufficiently heated and clean to insure the joint being quickly made, employed. The joint should then be wiped clean and allowed to cool. It must not be chilled or dipped in water.

With iron wire there is not this need of care in overheating, nor in slightly chilling the joint when made; the packing may therefore be abandoned.

The abandonment of the practice of bending the line wire at right angles after the joint has been whipped has a further advantage: the joint is far less likely in rough weather to become entangled with other wires.

It is well known that creosote is inimical to gutta-percha. All gutta-percha wire should therefore be kept apart from creosoted poles, or timber of any kind that has been so served.

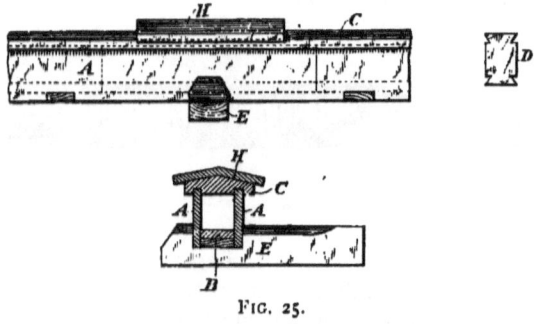

FIG. 25.

When wires insulated with percha are laid in boxing or casing, it is well to serve the outside of the boxing with a coat of Stockholm tar, but to leave the inside untouched. Carbolinium, a new preservative for timber, and one which can be laid on with a brush, should in like manner be confined to the outside of the casing.

Fig. 25 represents a form of casing suitable for tunnel work. It has the advantage that it is capable of being constructed without the use of a nail or screw. It is formed of four pieces of wood, the side pieces A A are grooved to accept the bottom board B; the top C is grooved to accept the side pieces; the two sides are held together by tie pieces D inserted at intervals of 3 feet. Instead of iron supports, wood blocks cut from flawed oak arms are shaped, as shown at E, to receive the boxing. These supports are fixed in the masonry by wood wedges or cement, 6 feet apart. A piece of wire passed around the support and the boxing holds the entire structure safely in position. No two pieces of wood of which the boxing is composed terminate at the same point, with the exception of the lid. At the junction of the lid pieces a wooden capping H is employed to cover the seam. Where boxing of a smaller section is required, it can usually be obtained cut from the solid. The joints should in this case be made obliquely, and care taken to avoid the joint of the capping taking place at that point which forms the junction of the other portion.

Wires insulated by gutta-percha should, where employed in signal boxes, be cased in; and where the wires are carried underneath the floors, the casing employed should be cut from the solid, and fixed in an inverted manner, i.e. upside down, so that any water which may leak through the flooring may pass away round the casing rather than penetrate it, as would be the case if it were fixed in the ordinary manner, face upwards. Where gutta-percha covered wires are exposed to such leakage, it has the effect of largely depreciating their life, due not merely to the varying conditions of moisture and dryness, but also to the soft-soap and soda used by the signalmen in washing the floors of their boxes.

All gutta-percha wire should be protected by a layer of ozokerite tape. The gutta-percha now generally used for the insulation of wires readily loses much of its insulating property if for long exposed to the air; and if wires insulated by gutta-percha unprotected by any covering are exposed to the air, it will be found that in three or four years the elasticity of the insulation will have disappeared, and that which remains will have become hard, and if bent, will crack, readily exposing the conducting wire within it. If the percha is covered with ozokerite tape, i.e. tape which has been saturated with ozokerite, and which is wound on to the wire warm, it will greatly lengthen the useful life of the wire. It would appear that in the manufacture of gutta-percha for telegraph purposes a great deal of oily matter is infused into it. This no doubt increases the insulating power of the composition, but that property is readily dissipated on the wire which is charged with it, and otherwise unprotected, being exposed to the air.

Gutta-percha is more readily affected by heat than is india-rubber. For this reason the author has for some years abandoned wires insulated by it for use on signal posts—wires required for electric signal repeaters, light indicators, &c. A similar gauge wire, say a No. 18 or No. 16, insulated with rubber, protected by woven or plaited hemp and compound preservative, similar to that employed for electric light wires,

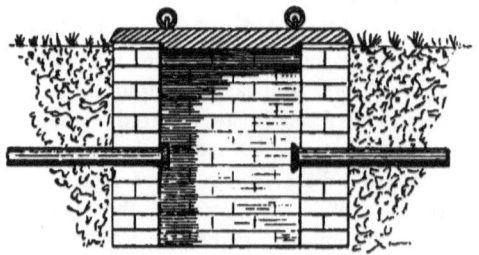

Fig. 26.

the wire being tinned, will prove much more durable. It is, indeed, questionable if india-rubber insulation will not to a great extent supersede gutta-percha, especially where joints in damp places can be avoided. With gutta-percha the ordinary lineman can make a fair joint, but he would require special apparatus and careful instruction to deal with vulcanised joints. Plain india-rubber would not present this difficulty.

All underground wires laid in pipes should be "drawn" (moved) periodically, so as to prevent them from becoming imbedded in any sedimentary matter which may be deposited in the pipes by water flowing through them. Flush boxes should be arranged at similar distances apart, so that the wires drawn out of any one section may be available for the adjoining section; and in all cases these flush boxes should be so constructed as to provide a well, Fig. 26, a depth of some two feet below the pipes. Any sediment which may then wash in from the surrounding soil will thus have an opportunity to settle in the well, and if this is cleared out occasionally, very little should pass into the pipes. The joints of all pipes should be sealed so as to prevent the ingress of water.

In station yards, where the wires are generally laid underground, and in other equally crowded places, armoured cables for odd lengths will probably be found not only useful, but, in order to avoid disturbing the body of through wires, a necessity. Such cables are extremely useful for crossing the line of railway for "repeater" or similar purposes, where, unless the wires are carried underground, higher poles are necessary in order to carry them overhead.

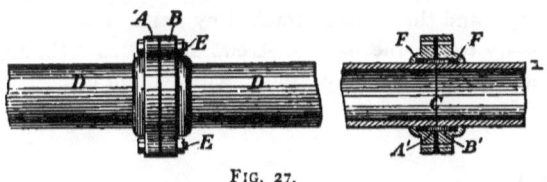

Fig. 27.

A simple form of cast-iron piping which, having an india-rubber joint, is very quickly placed in position, is shown in Fig. 27. It is inexpensive and readily handled. A, B are clamps which slide over D, which is merely a straight tube. A, B are recessed at A', B' so as to receive the small piece of tubing C, together with an india-rubber washer F at either end. The pipes D are laid out so as to meet at C. A and B, as well as C, are then slipped on, the tube C is brought into its proper position, and A and B are then closed up and tightened by the bolts E. The india-rubber washers afford a firm grip, so that the pipes D are not readily drawn apart. The arrangement is better suited for level ground than for heavy slopes. In the latter case it is preferable to employ pipes which admit of the joints being leaded, or if earthenware pipes are employed, cemented.

Wherever boxing can be used instead of pipes, it is in many ways preferable. It is easier of access, avoids interference with permanent way, and defective wires can be removed from it, or additions made, without trouble. The boxing does not become choked with sedimentary matter, and there is not, should a fault occur in any one wire, so long as the boxing remains water-tight, the same liability to stoppage as is the case with iron pipes—there is not the ready means of finding earth. There is, with wooden boxing, also less need, if indeed any need for lightning protectors, and where the use of these can be avoided it is desirable, as they are occasionally a source of interruption.

On railways it is often possible to support boxing on short standards. The boxing should not be less than a foot from the ground; but of course where a wall can be taken advantage of it is better to fix the boxing to it. If iron supports are used, and the boxing attached by nails or screws, care is necessary to see the nails or screws do not pass through the casing, and so cause injury to the wires within it.

to Railway Working. 59

CHAPTER IV.

TELEGRAPH INSTRUMENTS AND BATTERIES.

ALMOST throughout the entire railway service, the *Single Needle*, with the "drop," or vertical handle, is that employed for message work. By some of the companies the Bright's Bell is frequently employed at the terminal, or transmitting station, of a circuit, in connection with what is otherwise a single-needle circuit, and, in some instances, to form direct working circuits, or circuits having a very limited number of instruments upon them.

In Ireland the *Sounder* is largely employed for railway message work.

The Sounder is also used to a larger extent than formerly, for direct working circuits on railways in Great Britain.

The *Telephone*, although scarcely a message instrument, being more strictly a *vivâ voce* means of communication, is largely used by the entire railway service.

The *Single Needle*, Fig. 28, is a well-known and exceedingly useful instrument for railway work. It requires little, if any, adjustment, and admits of almost any number being joined up on the same circuit. The movement of the outer or indicating needle, Fig. 29, is contingent upon that of the induced (inner) needle, the two being arranged upon the same axle, the axes of which are enveloped by the similar poles of the two horse-shoe magnets A, B. This axle is formed of two parts soldered together at its centre. Each part becomes inductively magnetised by the magnets A, B. The two like poles adjacent to one axis impart magnetic properties to that portion of the axle, resulting in a south polarity of the inner needle at S, while the two opposite poles of the same magnets induce a north polarity in the upper arm of the needle at N.

It is undoubtedly desirable a limitation should be placed upon the number of instruments which should form a circuit. The conditions of the various services, however, differ to such an extent that it is difficult to lay down any rule upon the subject. The governing factors would appear to be: (1) the

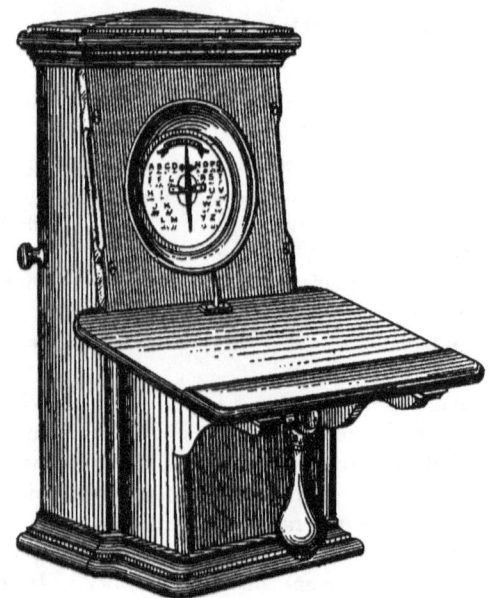

FIG. 28.

amount of work to be disposed of; (2) the stations, or signal posts, which require to be in direct communication.

The single-needle instrument is largely used for train-reporting circuits. Twelve to fifteen, and in some instances even more stations, or signal posts, are incorporated upon the same wire. Where the train reports are numerous, it is well to arrange for an up and a down circuit. It is then possible to deal with the reports with despatch; otherwise, should an up and a down train be reported at the same time, manifestly one set of reports must wait till the other has been disposed of.

It has been the practice of the author for many years past to employ an acoustic instrument, such as the Bright's Bell, at the transmitting station of all busy single-needle circuits. Even with the single needle, telegraphists now read much more by sound than by sight. The single needle is not, strictly speaking, an acoustic instrument, but the better insulation of to-day, as compared with that of some few years back, combined with the present improved battery service, enables the signals to be more forcibly rendered, and thus more capable of being read by sound. The Bright's

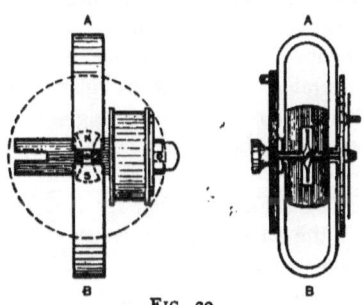

FIG. 29.

Bell is essentially an acoustic instrument, and the skilled clerk receives from it as he would from the voice of a person inditing to him the subject which he is committing to paper. The instrument, in fact, indites its communication to the writer. There is not that stress upon the clerk which attends the reception of messages by the movement of the indicator of a "single-needle" instrument. He is, as it were, employing but one of his senses—his sense of hearing. With the single needle, however experienced he may be, he cannot thoroughly rely upon the beats of the needle; he must from time to time call into requisition the faculty of sight.

The amount of telegraph work which now has to be dealt with at all the large centres of a railway system, imposes upon the staff uninterrupted labour. In many offices, clerks, on taking duty, are occupied without intermission for hours together, always working under stress, and any amelioration that can be afforded them is in every way desirable. The employment of the "Bright's Bell" as suggested will effect this.

It will be inferred from what has been said that the author's views favour the employment of the bell rather than that of the single needle. That is so in respect of those circuits, limited to four or five instruments, where speed in transmission is required, and at busy stations on single-needle circuits. The bell instrument is not so simple nor so easily managed as is the needle instrument. It involves the employment of a relay. There is from time to time adjustment of this relay and other parts; it is not therefore desirable it should be placed in the hands of junior or inexperienced clerks, or signalmen. Hence the propriety of retaining the single needle for minor stations as well as signal boxes.

Although the sounder is used on the Irish railways much as the single needle is on the English lines, it has never taken root with the latter, and where employed, it is generally used for direct working between important centres, either with or without a relay.

On some few lines duplex working has been adopted, and in one or two instances quadruplex has been found useful. There is no doubt that both duplex and quadruplex working are destined to be of considerable service to those companies whose needs call for heavy and continuous telegraphic correspondence between important centres. A duplex circuit will not, as a rule, dispose of work so quickly or so conveniently as two simple circuits. The cost of working will be about the same, and obviously two wires are safer than one, for if that which is duplexed breaks down, it is tantamount to the stoppage of two wires, or nearly so. It is improbable with two working wires, both will be interrupted at the same time. Where the distance is not great, and the cost of providing an additional wire will not be excessive, and other means of communication between the two points is restricted, it is better to afford the additional wire. Where the distance is considerable, economy will attend the establishment of duplex working, so arranged that the apparatus may be used either *simple* or *duplex* as required. This can be effected by a switch usually obtainable with the duplex apparatus; and where the work to be dealt with is greater during any one portion of the twenty-four hours, nothing is easier than to arrange the staff so that the requisite number of telegraphists may be in attendance to meet the duplex requirements.

For distances of say one hundred miles and over, double-current keys with relays will be found desirable. With the positive and negative currents following in succession, the signals are rendered with greater precision than is the case with the single-current key.

Perhaps no electrical instrument, apart from the Block instruments themselves, has proved of greater service in aiding the working of railways than the *telephone*. The main lines of most leading railways are now fully equipped with a telephone service, by means of which the signalmen throughout the route are enabled to speak to the neighbouring boxes on either side. As a rule the circuits should be limited to ten instruments, and where the communications are numerous or immediate attention is requisite, the number of instruments should be still more restricted. In the neighbourhood of junctions, or where "lie-by" sidings exist, especially in foggy weather, the value of such a communication between the different signal boxes cannot be overestimated. By its means the signalmen are enabled to learn from one another what trains, if any, are standing, and to afford despatch to those of the most urgent character. In this alone, at many crowded centres, the cost of the communication has been repaid many times over.

Although telephonic communication has been thus almost lavishly provided, and although it is in every way taken full advantage of, it has, in very few instances, superseded the ordinary telegraph instrument. Where such has been the case, the latter have been laid down and used purely as a local communication. In such cases, generally the telephone is the better form of communication, and as such it is frequently used for announcing to signal boxes, by the telegraph office, the running of trains ; and for collecting from the signal boxes the time of departure of such trains as have to be signalled elsewhere.

The telephone is a *vivâ voce* instrument. Its design was that it should enable conversation to be carried on between points apart from one another. There is no reason why messages should not be transmitted by its aid, and written down for delivery the same as is done in the ordinary telegram ; and where a communication has to be conveyed to any other than the person to whom the subject is being spoken, it is preferable it should be sent as a message, and if of importance repeated by the receiver back to the person by whom it is uttered. Although in every respect a marvellous instrument, words uttered by it are liable to be misunderstood. Much depends upon the voice of the speaker, and experience in the use of the instrument.

As a check against omissions where trains have to be announced to, or their departure reported from a signal box, it is a good plan to employ check numbers. Each announcement is thus accompanied by a number, which should be recorded with the details of the train report, and be available for reference in case of dispute between those responsible for its transmission or receipt.

Where a telephone circuit has to be erected upon poles carrying other wires, the telephone circuit should be formed of two wires—a metallic circuit, well insulated, so as to be as free as possible from " earth." Copper wire should be used, and the wires should revolve around each other, the revolutions being completed as nearly as can be at equal distances. Where two wires only are concerned, unless caused to revolve with other two wires, there will be a waste of space. It may in such a case be as well, provided the circuit is not a very long one, to meet the induction arising from neighbouring wires by running the wires required for the telephone circuit " parallel," and crossing them at given distances. To carry this out effectually, it is necessary to pay strict regard to the influence of each neighbouring wire. Assume that one of the adjacent wires accompanies the telephone circuit throughout. Its disturbing influence may in that case be regarded as practically the same at all points. If this is so, then we may counteract its effects by cutting the telephone wires, and crossing them, midway. This should, to all intents, cancel the disturbance derived from that wire; but it is clear we may have other wires above or below one or the other of the telephone wires, or both, and they may not all accompany the telephone wires throughout their length. Here the careful observation and ingenuity of the engineering officer will avail much. He must deal with the difficulty piecemeal; consider each wire, commencing with the shortest; cross the telephone wires to counteract its influence; then consider the next longer wire, and cross again to meet it; and so on. Obviously, however, this will not in all cases admit of the crossed sections being equal in extent; so that, on the whole, it may be found that crossing the wires at every fourth or fifth pole will prove the safest and most practical mode of dealing with the difficulty. What we have to bear in mind is that the sections of wire between each point, at which they are crossed, should be equal, and that the number of crosses inserted should be such as to render the counterbalancing effect equal. Thus, a length of line divided into two equal parts, the disturbing influence being the same, crossing the two telephone wires midway should eliminate the inductive disturbance; but if the length were divided into three equal parts the result would not be satisfactory, for the sections would not counterbalance one another: the proportion would be as 1 to 2; that is, there would be two sections of the telephone circuit subject to induction in one direction, and one section only to induction in the opposite direction. Practically this would mean one section unprovided for, the inductive influence upon which would still be found extremely objectionable.

Where more than one telephone circuit has to be run upon the same line of poles, it is better where the wires are or can be erected four on an arm, to erect them on the "revolving" principle. If the wires are carefully erected, the arms being

F

one foot apart centre to centre, and the insulators a similar distance, the four wires on one side of the pole will at each post form a square, and this square, although of a smaller section in the middle of the "bay," will, if the wires are properly erected, be maintained throughout. Wires thus erected complete their revolution once in every four "spans," and thus counteract by the change in their relative positions any inductive effects from other wires than those which run parallel to them on the same poles.

Although perhaps scarcely within the scope of this book, it may be desirable to point out that where more than one group of telephone wires erected on this principle occupy the same poles, there will be "parallelism" between the wires of the several groups or squares, and what is termed "overhearing" will arise. Conversation passing on one pair of wires will be heard on the pair of wires parallel to them. Thus in Fig. 30 the pair of wires 1, 3, which we may take as forming one circuit, will be parallel with wires 5, 7, forming another circuit, and wires 2, 4 will be parallel with wires 6, 8; and, consequently, what passes on the one pair will be heard on the other. Here the difficulty is to be met by crossing the wires of one group at points equidistant one from the other. It will, in the case under review, be clear that all things being equal, the case may be met by crossing the wires once only —midway of the entire length: but this assumes all things equal—equality in gauge of wire, length of circuits, leakage and other influences—and this condition is seldom, if ever, attained; so that in practice it is found preferable for long circuits to make the crosses at distances not exceeding ten miles.

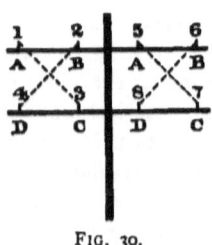

FIG. 30.

Where the wires are thus run on the revolving principle, and crosses have to be inserted, it was originally the practice to terminate and cross-connect the wires; but the desired result may be achieved by, at the given points, running or cutting the line wires, and so connecting them that for one span they may take a *backward* twist. Thus, if the twist adopted is a "right-hand" twist, all that is required is to give the wires to be crossed a "left-hand" twist for one "span." Let the arrangement shown in Fig. 31 represent the squares formed by four wires on a line of poles, and the letters the position which each wire takes in its course as it is made to revolve around the other wires. At the second pole A, it will be observed, takes position 2, B changes to position 3, C to position 4, and D to position 1. A and C form one pair of wires, serving one circuit. It is necessary, in order to counteract inductive influence, that these wires should be crossed. Ordinarily at the third pole A should occupy position 3, but instead of doing this we give all four wires for this span a backward

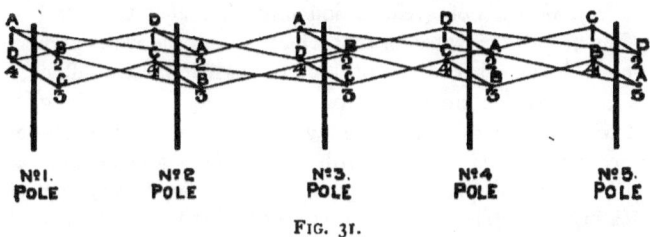

FIG. 31.

twist, and thus A, instead of taking up position 3, is put back to position 1, and all other wires in a similar manner, instead of taking one step forward take one step back. It will be seen this does exactly what would be done were the wires terminated and crossed by cross-connecting pieces of wire. At the next pole the wires again pursue the right-hand twist.

With a crowded line of telephone wires it is not quite clear to what extent this "crossing" can be carried out with effect. One circuit of course remains intact. That naturally would be the longest circuit. The next most important would require crosses sufficient to counteract the induction from the first named circuit. A third circuit would probably require double the number of crosses inserted in number two circuit. Number four circuit would again require double the number imparted to number three ; but whether, after having thus dealt with several of the circuits, it would not then be found that the inductive influences had been so disturbed, as, if not to entirely destroy one another, at all events to obliterate articulation, we may go back again to a group of wires without crosses, and, as it were, make a fresh start, is a question which must be solved by experience. No doubt the induction between a large number of circuits on the same poles will tend, when the circuits are all in use, to destroy overhearing, but there would still remain a *disturbance* on the wires which would not be conducive to articulate speech. Whether the question of induction will, in respect of trunk lines of important wires, not prove a greater factor than the capacity of the poles in determining the number of wires which a line may carry, is somewhat a question of the future.

The position of greatest inductive influence between two wires or pairs of wires, is that which places the wires parallel one with the other. The position of least inductive influence is that in which the wires cross one another at right angles. It will be clear that, if desired, wires for telephonic purposes may be erected parallel with other wires, and crossed at given points, as previously explained. So far, this system of dealing with telephone wires has been limited to a pair of wires ; but, manifestly, to place it on a level with the revolving system it is necessary it should be applied to four wires, and that these four wires should be crossed at given points dependent upon the influence derived from neighbouring wires, so as to change their relative positions in the same manner as wires run on the revolving principle. The result would be the same as that ascertained by pursuing the revolving method, the wires on the poles preserving their parallel appearance. It may be that no advantage is, in point of working, secured by the adoption of this method, nor, apparently, is such the case ; but it may be of service in cases where telephone wires are required to be erected on poles carrying wires for other purposes, and which are erected on the parallel system. Revolving wires erected upon poles carrying wires erected on the parallel system are not conducive to uniformity, or the ready detection of a deranged wire.

Where the leading-in wires of more than one telephone circuit are carried within the same casing, or in the same pipe, it is of the utmost importance that the wires should be twisted together so as to revolve around one another, as has been previously indicated in respect of open wires. A few feet of covered wire in close proximity to other wires, is naturally

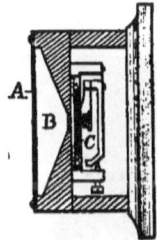

Fig. 32.

more subject to disturbance from such wires than is the case with open wires one foot apart; and where two telephone circuits are thus in juxtaposition, and parallel even for a few feet, overhearing will be possible.

It has been asserted that the telephone is a disseminator of

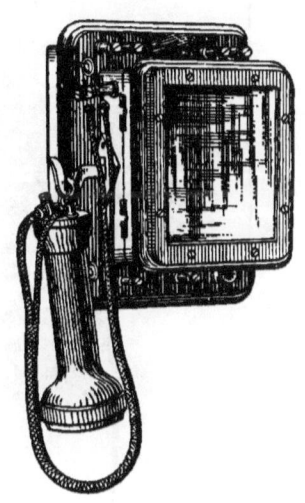

Fig. 33.

disease; that where used by a person suffering from internal diseases, the germs of the disease may be deposited upon or within the mouthpiece, and inhaled by any person using the instrument within a reasonable time thereafter. This has been met by a recent invention, as shown in Figs. 32 and 33, in which the ordinary mouthpiece is replaced by a diaphragm of wood or metal, which affords a plane surface free from interstices in which anything of a deleterious character may lodge. The ordinary mouthpiece being removed, an outer diaphragm A, Fig. 32, is superimposed upon the inner one B, so as to inclose a film of air. On speaking to the outer diaphragm, or in its neighbourhood, it responds to the vibrations of the voice, and by means of the enclosed air conveys those vibra-

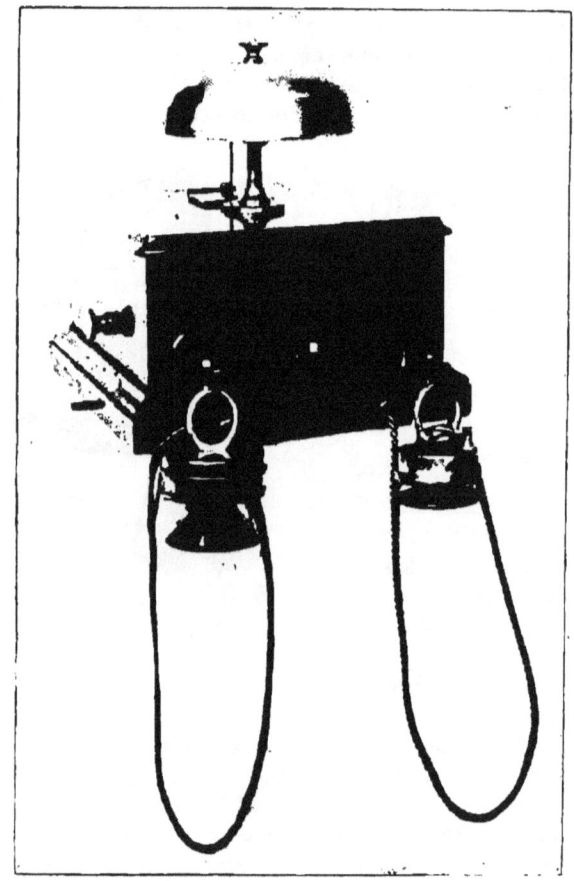

FIG. 34.

tions to the metal diaphragm C, which causes the required intonations to be transmitted on the wire. It might be expected that the voice would under this arrangement lose much of its force; such, however, is not the case. There is apparent to the specialist a slight diminution in power, but it is so slight as in no way to interfere with speech on long distance lines; while the clearness of the voice is even enhanced.

On the Midland Railway the author has, with great advantage to the traffic, employed a *combined Block Bell and Telephone* at such points as, from a traffic sense, would admit of it. The same wire serves the telephone. The block bell becomes the telephone bell for the purpose for which it is required—to call attention. A call signal, indicating that the signalman's attention is required on the telephone, should be added to the Bell code.

This combination necessitates the application to the ordinary block bell of those parts of the telephone requisite to carry on conversation: a switch for switching the bell or telephone in circuit, together with the usual *receiver* and *transmitter*. These can, with perfect convenience, be combined directly with the block bell, or connected with it electrically by means of a pillar switch. Both are illustrated by the accompanying Figs. 34 and 35.

FIG. 35.

The Application of Electricity

The pillar switch has been provided in order that the application of the telephone may be effected without disturbing the bell beyond the circuit connections, which are shown in Fig. 36. It is perhaps unnecessary to add that the small form of *transmitter* and *receiver*, shown in the illustration, is preferable for use in this combination, as occupying less space and being more convenient for ready handling by the signalman.

The economy effected by the adoption of this arrangement rests mainly in the saving of the line wire, which otherwise would be required for the telephone. Having regard to the

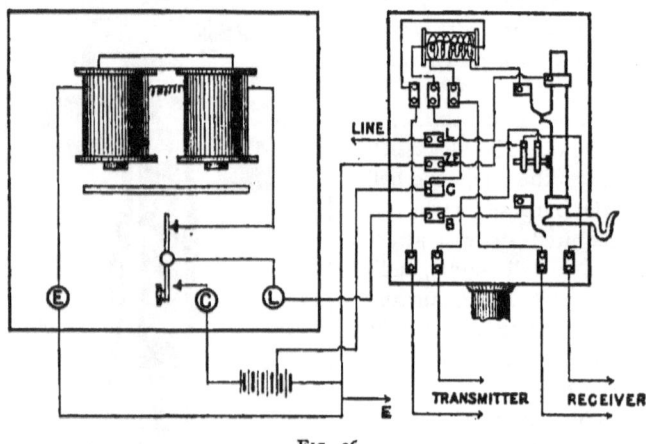

FIG. 36.

convenience which attends the ready means of communication thus afforded signalmen, and the aid which such a mode of communication between signalmen renders the traffic, there should, now that telephonic apparatus is so cheap, be no hesitation in affording every block signal post not provided with a more independent form of telephonic communication, with this simple and economical adjunct. Where the traffic is heavy, and the signalling between block post and block post practically continuous, of course telephonic communication must be carried on by means of wires independent of the bell wire, but where the trains are not numerous, and where the communication can be restricted to the block section, its utility must commend itself to all railway men.

Under no circumstances should the telephone, or in fact any form of communication, be allowed to supersede the use of the block signalling apparatus, or in any degree to interfere with the working of the latter.

The Phonopore is an instrument operated by induced currents. Its transmitting portion consists of an induction coil, the primary of which is in circuit with the sending key and battery, while the secondary is connected with the line wire on one side, and with what is termed a *phonopore* on the other side. The phonopore (or *sound passage* as it is termed by the inventor) is in point of fact a small helix composed of two wires of similar gauge and capacity. The two opposite ends

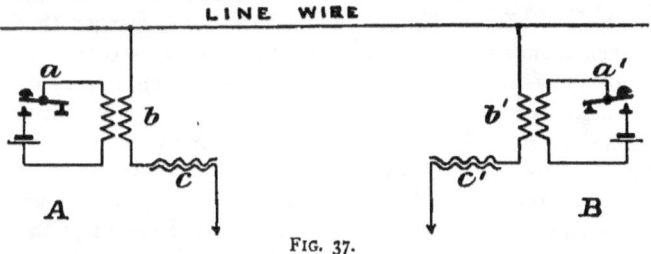

FIG. 37.

of these coils are left disconnected. Induction takes place between the two coils to a sufficient extent to admit of the formation of working signals. The foregoing diagram, Fig. 37, will perhaps serve to make the arrangement, in principle, clear to the reader.

Assume the line wire to be that applicable to any circuit, say a wire worked duplex between Derby and St. Pancras, London, as is the case, and that upon it, serving two other stations, there is imposed phonoporic apparatus as shown at A and B. On pressing down the key a, the local circuit through the primary of the induction coil is formed, and the armature in circuit therewith is set in motion. By the adjustment of this armature we may get vibrations of any amplitude, and so set up corresponding currents in the secondary coil b. These induced currents will be, necessarily for the purpose in view, of high frequency, and alternating in polarity, and will, consequently, not affect the working of an ordinary telegraph instrument operated by direct currents. From b the induced currents will pass, in one direction, to the line wire, in the other direction, by induction, to earth. Arrived at B, they will pass through b', and the phonoporic coil c' to earth ; and if a suitable receiving instrument is inserted, the signals formed at A, by the manipulation of a, will be read off at B. The signals usually employed are the Morse alphabet. A telephone may be used for a receiver, but a proper instrument, rendering the elements of the letters on a sounder or printer by means of a relay, may with advantage be employed.

The introduction of these instruments on an ordinary telegraph circuit, or indeed their employment on any wire, has one baneful effect. The induction effects are so marked that all neighbouring wires are affected, and telephonic communication more or less disturbed. In course of time telephonic operators become used to the disturbance on the telephone due to this cause, but to a novice it is extremely troublesome. The effect is so great that messages signalled by means of the phonopore have been read off upon telephone wires, which have traversed but a slight section of the route of the phonopore wire, at stations many miles distant.

With the object of obviating this by intermingling the inductive effects on other wires, the author has employed two phonopores at the same station, but the *pitch* of the vibrating reeds not harmonising strictly one with the other, the effort has not proved successful ; the expert is able to eliminate the one signal from the other, and by fixing his attention upon that note which he is desirous of reading, can follow the communications with perfect ease. Although it should, of course, be quite possible to produce vibrating reeds of the same pitch, it would appear that considerable difficulty attends their employment for the purpose under review, in that condition and under those circumstances which would enable the effect produced by the one instrument upon neighbouring wires to be counteracted by that resulting from another instrument. Nor is it by any means clear that, should success attend any such effort, it would not be accompanied with disastrous consequences to the phonoporic communication itself; that the cross induction between the signals passing upon the two circuits would not so affect the signals as to render each of them unintelligible. To hope that the evil effect of this induction may be overcome is apparently to hope for too much. We must not lose sight of the fact that it is the basis of the application ; that to produce the responsive signal in the phonopore, currents of a certain potential have to be generated, and to reduce their inductive effect the potential of those currents must be reduced which, reason would teach us to believe, would be prejudicial to the mode of working. Needless, perhaps, to say, if it were possible to overcome it, the invention would be a most useful one, and would largely extend the capacity of all telegraph wires.

Condensers would, of course, answer the purpose of the phonoporic coils c, c'.

BATTERIES.

The *Daniell* form of battery has of late years been largely superseded by the *Bichromate* and *Leclanché*. The Bichromate is an exceedingly reliable battery for constant-current purposes, and where a number of instruments are required to be worked off one set of batteries it stands unrivalled. The Leclanché is a most useful and economical battery—one which, if judiciously applied, is amenable to most requirements of a railway telegraph service. It possesses a great advantage, in that so long as the liquid is saturated with the salts, it does not freeze under the temperature of an ordinary winter such as is experienced in this country. It readily polarises, and as readily, when set free from action, depolarises. It is practically free from waste of material when not in operation. To use it with advantage, however, for constant-current purposes, such as signal repeaters or for block working, the porous pots require to be made in such a manner as will obviate their ready destruction by the formation of crystals within the pores of the ware. It is now many years since the author began to substitute the Leclanché for the Daniell cell. The difficulty which he then had to contend with was the freezing of the latter during prolonged frosts. Experiment showed that the Leclanché withstood the temperature of our winters without materially detracting from its value as a battery. It was clear that such a battery would obviate much difficulty, but it had never been regarded as capable of sustaining a constant current; added to which was the difficulty of the ready rupture of the porous pot. Discussing the matter with Mr. John Fuller, of the firm of Messrs. Fuller and Son, of Bow, London, Mr. Fuller soon placed at the author's disposal a Leclanché not only capable of sustaining the demand for a constant current, but one, the porous pots of which did not readily burst or become disrupted when so used.

It is the practice of the author to employ two classes of Leclanché : one for intermittent currents, such as block bells or message instruments, the other for block working, signal and light repeaters, requiring a constant current. The resistance of the cells employed for the latter purpose is higher than that employed for the former work.

In the replacement of the Daniell by the Leclanché battery, a very large saving may be effected. Not only does the latter avoid all the labour and destruction attendant upon thawing frozen batteries, but it is in itself a more economical battery—less costly to maintain, far more durable, and may be placed in any convenient position beneath the signal box, or even in closets outside it.

It will be found advantageous to use the Bichromate battery for heavily worked message circuits, and especially for "compound" working, where several circuits are worked off the same set of batteries.

For repeaters, the No. 1 Leclanché cell is most economical. One cell for the ON, and one for the OFF indication.

Light indicators worked on the principle adopted by the author, affording a continuous ringing of the bell when the indication shows the light to be OUT, will require two No. 1 cells.

For block working, four No. 2 cells are usually employed, although less would meet the demand.

The electromotive force of the Bichromate cell is, approximately, 2·0; that of the Leclanché, 1·6 volt per cell. There is no great variation in the output of the Bichromate cell when called into constant action, but the force of the latter falls considerably immediately it is brought into work. Polarisation, however, soon reaches its limit, and thus reduced it remains pretty constant. It is here that it becomes useful to meet the constancy of current which block working and repeaters demand. It is necessary to allow a margin for this rapid fall, after which the cell will be found to do its work well.

The most recent introduction in batteries is the so called "dry cell." These cells require to be kept in a moderately dry position. When in stock, and of course equally so when in use, they should not be allowed to rest on their side, as, under such a condition they, becoming abnormally dry on one side, lose their power. Although an exceedingly convenient form of battery for many purposes, it is questionable if their employment for railway telegraph purposes can be attended with any great advantage. The condition of the battery is not observable externally, it can only be ascertained by testing. When failure arises the cell must be replaced. Having regard to the fact that wherever block signalling is in operation a lineman must be within easy reach of any defect which may arise, the employment of the dry cell would not, apparently, be attended with economy in labour, even were it shown that the cells were, as compared with a Leclanché cell of similar power, less costly cell for cell. With the ordinary Leclanché, the lineman knows generally from its appearance what its condition is. He is, to all intents and purposes, aware when the zinc will require replacing, and if it should be partially consumed, the cell can still be maintained by lowering the upper portion of the zinc rod into the electrolyte. The only portion which is not exposed to visual inspection is the negative element. It is not possible to tell to what extent the binoxide of manganese has been reduced, but, as is known to all who have used this class of battery, the depreciation in this is so slow that it affords ample time for replacement before absolute failure occurs.

Messrs. Siemens have recently introduced a new salt—*Siebrosal*—which they propose should take the place of salammoniac, and which it is stated will not freeze at temperatures down to 12° Fahr. (20° of frost). The electrolyte then assumes a gelatinous condition, without, however, it is said, affecting the efficiency of the cell. This salt, like salammoniac, is deliquescent.

The following is the result of a test made with the object of showing the fall in E.M.F. of an ordinary Leclanché cell and a "dry" cell, each cell being short-circuited for the time shown through a resistance of 100$^\omega$.

THE OBSERVED FALL IN POTENTIAL OF AN ORDINARY LECLANCHÉ AND A "DRY" CELL, EACH BEING MAINTAINED CONTINUOUSLY IN CIRCUIT WITH 100$^\omega$ RESISTANCE.

E.M.F. Leclanché Cell. (Volts.)	Period of Observation. (Minutes.)	E.M.F. "Dry" Cell. (Volts.)
	After.	
1·35	..	1·40
1·25	15	1·38
1·24	30	1·38
1·23	45	1·38
1·23	60	1·38
1·23	75	1·38
1·24	90	1·37
1·24	120	1·37
1·23	180	1·36
1·18	660	1·35

CHAPTER V.

BLOCK SIGNALLING.

DEFINITION OF TERMS.

THE expression "block system" may be defined as—

A method by which the traffic on a line of railway is so regulated that one train or engine only shall be in any one section, and upon the same line of metals, at the same time.

To effect this it is necessary to divide the line into lengths, or sections. Each such length is termed a *block section*.

Block signalling has been conducted under three different denominations or methods of working—

(i) The *positive*, under which the section is maintained blocked during the time a train is in the section, or the section is fouled by shunting operations. The indication at other times being that of *all clear*.

(ii) The *affirmative*, under which the signals are normally at *line blocked*, and are, on request, when the section is free, placed at *all clear* for the admission of a train. A system which requires that before a train is sent forward, permission to do so shall be asked of the station in advance.

(iii) The *permissive* system, under which two or more trains travelling in the same direction are allowed within the same section at the same time; the second or following trains being cautioned as they pass the signal box that there is a train within the section in advance of them.

That the latter system is a pernicious system has now been fully recognised, and it may be said that it is now employed only under exceptional circumstances, such as station yard working, where trains are required to move with caution, in order that they may be brought to a stand at their respective platforms for interchange of traffic or connection with other trains.

The *affirmative* system is now that generally employed. The *positive* system would appear to commend itself for adoption, as indicating more truly the real condition of the section, and as calling for less signalling than that under which the signalman is required to ask permission to send forward a train. With the single-needle block instrument, which is now so employed that it affords three indications, viz. " Line clear," " Line blocked," and " Train on line," there is not, it is true, that inconsistency which attaches to other apparatus employed under this system, and which limit the indications to " Line blocked " and " Line clear." There can be no question that the block apparatus should accurately represent the condition of the line. It is very questionable if a greater degree of security is attained, as in some quarters is assumed to be the case, by maintaining the line blocked at all times except when cleared to admit a train. That the art of block signalling has much advanced of late years no one will question, and the extension of the "affirmative" system is one of the points to be noted in connection with this advancement. The human machine is by no means infallible, and the question arises: Is the signalman less likely to commit an error when he is required to work his signals solely in accordance with the movement of the trains—i.e. with the fouling or clearance of the section—than when he knows the train last sent into the section has passed out of it, but yet his block instrument indicates *line blocked.*

It may be asked: Is the importance of a signal emphasised or otherwise by being pulled off in front of an approaching train? Does the knowledge that the system provides for the signals being maintained at danger when there is a possibility that danger is at the time non-existent, impress those in charge of trains with that need for strict observance of the signal that would attach to it if it were *known* to be an indication of absolute danger?

It is quite clear that the *affirmative* system calls for more signalling than the *positive* system. This is not possibly a very material point, still a multiplicity of signals is a mistake. The signalling should be reduced within the narrowest limits compatible with the absolute needs of the service. Every signal in all probability calls for the movement of the signalman from one position to another. In a busy box this constant movement to and fro, from one end of his signalling shelf to another, and still worse, where the instruments are not fixed over his signalling frame, between the latter and the position of the former, results in so much exhaustion of the human frame; and where this can be reduced by a reduction in the number of signals, it is obviously desirable. It is in many instances undoubtedly of advantage to know the description of the train which is approaching, but whether this might not frequently be abandoned is a question which may well demand consideration at the hands of the traffic or other department of the service responsible for drafting the signalling instructions.

Consider a line filled with trains, similar to the Metropolitan! How would the application of the affirmative system affect such a traffic? It may be argued, and no doubt with weight, that there is a want of analogy between the traffic of such a line and that of railway lines in general; that there is an absence of junctions, and that danger attends the employment of the positive system at junctions. But where the positive system is employed, necessarily the junctions must be blocked either constantly, and the block only removed when required for an approaching train; or the signals for conflicting roads must be placed at danger before allowing such roads to be fouled by an approaching train. There can be little doubt that in such a case as that quoted, the employment of the "affirmative" system would be attended with no advantage. Will the traffic on other lines, now worked under the "affirmative" system, grow to such an extent as to assimilate in character with that of the Metropolitan; and if so, will the "positive" system take the place of that now in use? If the traffic on the Metropolitan renders it necessary, and the traffic on other systems should grow to equal dimensions, it would appear safe to infer that the latter would eventually be called upon to employ the "positive" system.

The object of the semaphore or other signal is to protect trains in advance of it. We may establish rules which shall render any disregard of the signal penal, but its value *as a rule* must be largely enhanced when it is recognised that it is, *de facto*, an indication of danger.

These arguments have to be discounted by the fact that, as stated, the greater portion of British railways are now worked under the "affirmative" system. The *pros* and *cons* have, no one will question, received that consideration which the importance of the subject demanded; and we are bound to receive the decision arrived at with that regard to which it is entitled. At the same time it is but right to place the question in all its bearings before those whom it may affect, either in the present or in the future. Broadly speaking, it is regarded that the "affirmative" system is that most suitable to all the circumstances of ordinary railway traffic; and the main reasons for this would appear to be, that it is desirable for junction working, and to avoid blocking back or "obstruction" signalling when fouling roads at stations for station working; and that by some the application for permission to send forward a train—resulting, if the line is clear, in the lowering of the danger signal—is regarded as emphasising the fact that *the section is clear*, and so contributing to greater safety. The reader will bear the several arguments in his mind. A careful consideration of them will be material in enabling him to adopt that which may prove of the greatest advantage for protecting that description of traffic to which it is his desire to apply it.

An important factor in the satisfactory working of traffic under the block system is the *length of the block section*. These sections have necessarily to be regulated to meet the traffic required to be passed over each in a given time. Obviously, with a system of manual-worked signals, with wayside sidings and stations to provide for, it is not possible to determine these sections merely by a given length of line. Each has to be considered in relation to its demands, and so arranged that the time which trains will occupy in disposing of their work and passing through it shall, as nearly as possible, assimilate with that of other sections carrying similar traffic.

The length of section will thus vary according to the traffic and the impediments which stand in the way of its rapid disposal, and consequently the signal boxes will not, as a rule, be within sight of each other. Here electrical apparatus becomes necessary, in order that the signalling of the trains may be effected in such a manner as shall enable the signalmen to carry out the principle of block working—that is, by preventing more than one train being in the section at the same time.

At the date of the issue of the author's previous work on 'The Application of Electricity to Railway Working,' block signalling was by no means generally adopted. It was in use on the larger railways to some extent only. Under the Regulation of Railways Act, 1889, the adoption of "the block" has been made obligatory, and the traffic of all lines worked by more than one engine is now worked under it in one form or another.

The principle of block working is now well known. The electric signals are employed to guide the signalman in the manipulation of the signals which are provided for the guidance of the engine driver, and which we may denominate, for purposes of distinction, the mechanical signals.

Some form of electrical acoustic signal—usually a bell—is necessary for the purpose of calling attention to, and passing certain signals between, signal post and signal post. Also a description of "block" signal for each line—a signal which shall indicate whether the line is clear or blocked, as may be required.

Instruments which meet these requirements, worked by means of *one wire* or by means of *three wires*, are employed.

Those worked upon *one wire* only are actuated by "momentary" currents, i.e. currents effected by a momentary or transitory contact between the battery and the line wire. The objection to this type of instrument presides in the fact that there exists the possibility of a false signal being rendered by atmospheric electricity (lightning), or by the wire serving the block instrument forming contact with another wire on which currents may be passing.

Such instruments can be arranged so that the indicating portion can be brought into play only on receipt of a given signal on the bell, but this, as will be understood, requires a special provision which tends to complicate its manipulation while it does not fully meet the objection, inasmuch as the moment the receiving portion is brought into circuit may be the very moment the wire may be affected prejudicially. Of course this is somewhat improbable — the chances are against such being the case, but the chance is there.

The "*three-wire*" system admits of the appropriation of one wire to the up road or up trains; one to the down road or down trains; and the third wire to a bell communication common to both. The term is not applicable to any peculiar description of instrument; any design of block signalling instrument may be employed. Its advantage is that one wire being appropriated to the protection of one road, it enables the signals "train on line" and "line clear" to be rendered by permanent or continuous currents; or, under any circumstances, that one of these or similar indications shall be produced by the presence of the current.

We have thus block signalling capable of being conducted upon a single wire and under a three-wire system. Of the former the most perfect are Preece's and Tyer's. The instruments of the former are largely used by the London and South Western Railway, and by other companies to a less extent; those of the latter by many of the companies.

The three-wire method may be said to be confined to the single-needle block, Preece's three-wire system, and to a slight extent to Tyer's three-wire block.

There is, however, yet another system of somewhat recent introduction, known as Tyer's *Tablet Block*, and which is really an electro-mechanical instrument. In competition with it is the invention of Messrs. Webb and Thompson of the London and North Western Railway, styled the *Electric Staff Block*, to *Railway Working.* 85

The purpose and object of each of these instruments is to facilitate, as well as to protect, the traffic on single lines of railway.

Under the tablet system the trains are controlled by a tablet which has engraved upon it in bold letters the section of line to which it applies, and which is so designed that it will fit those instruments only which belong to that section. In like manner the electric staff applies to, and is available for, the instruments of one section of line only. The electrical duty of each is the operation of certain parts which admit of the extraction of a tablet, or a staff, as the case may be, and the principle upon which they work is that only one staff, or one tablet, can be out at one time. If no train is allowed to traverse a section without this emblem of authority, and as the emblem must be out of the instrument to allow it to be used, and if only one emblem can be out at one time, it is clear that one train only can be in the same section at the same time.

The following statement indicates the various block systems in use by the chief railways.

BLOCK TELEGRAPH SYSTEMS IN OPERATION ON THE FOLLOWING RAILWAYS.

Railway Company.	Systems in Use.	Interlocking.
Caledonian.	Tyer's one-arm three-plunger semaphore. Train tablet.	In operation on one short section covered by six signal posts, but presently to be extended to about fifty signal cabins on new lines in and around Glasgow.
Glasgow and South Western.	Tyer's one-wire.	None in use.
Great Eastern.	Tyer's single line one-wire block. Tyer's double-line one-wire block. Tablet block for single lines. Single-needle block (Great Northern form) on Great Eastern and Great Northern joint line.	Sykes' electric lock and block, with rail contacts and electric bars. (Metropolitan and suburban district.)

BLOCK TELEGRAPH SYSTEMS IN OPERATION ON RAILWAYS—
continued.

Railway Company.	Systems in Use.	Interlocking.
Great Northern of Ireland.	Harper's. Preece's. Electric staff.	None employed.
Great Western.	Spagnoletti disc. Webb and Thompson's electric staff. Tyer's tablet. Tyer's block. Preece's block. Single-needle block.	To a slight extent.
Great Northern.	Single-needle block.	To a slight extent.
London and North Western.	Three-wire. Tyer's one-wire. Electric staff. Two-wire (L. & N.W.) Three-wire (Midland).	None employed.
London and South Western.	Preece's three-wire. Preece's one-wire. Tyer's tablets.	Three-wire block and interlocking with mechanical signals combined. Tyer's tablets and interlocking with mechanical signals combined.
London, Brighton and South Coast.	Tyer's. Webb and Thompson's.	Sykes'. Saxby and Farmer's.
London, Chatham and Dover.	Single needles with bells attached for describing trains only. Sykes' double-arm bell blocks.	Sykes' electric locking gear in use all over the railway and branches. Also 78 treadles, mostly fitted to advance signals, and 61 fouling bars used for protection of junctions and cross-over roads. 57 electric shunting signals, chiefly at Victoria Station. 64 electric facing point detectors.
Lancashire and Yorkshire.	Tyer's one-wire. L. & Y. three-wire. Tyer's train tablets.	Only in use experimentally.

BLOCK TELEGRAPH SYSTEMS IN OPERATION ON RAILWAYS—*continued.*

Railway Company.	Systems in Use.	Interlocking.
North British.	Three-wire single-needle block. Tyer's ordinary block. Tyer's train tablet. Electric train staff.	Not yet introduced.
Midland and Great Western of Ireland.	Webb and Thompson's electric train staff. Harper's. Preece's.	None employed.
Midland.	Needle three-wire block. Tyer's tablet. Webb and Thompson's electric train staff.	Some 25 miles three-wire single-needle, interlocked with mechanical signals.
North Eastern.	Single-needle block, with the three usual indications. 'Walker' semaphore block instruments are used for working certain mineral lines.	Not employed except for a particular purpose for the section upon each side of the cabin upon swing bridge at Selby.
South Eastern.	One-wire system. Walker's bell, semaphore and ringing key, and Leonard's train in section disc. Leonard's commutator and disc added. Tyer's. Harper's.	Sykes' instruments and treadles employed between Cannon Street and Charing Cross stations.
Manchester, Sheffield and Lincoln.	Hampson's three-wire block. Tyer's three-wire block. Tyer's one-wire block.	Ross, Wharmby and Hampson's lock and block. Sykes' lock and block. Evans' lock and block.
Great Western.	Disc block. Electric train staff. Electric train tablet. Tyer's block. Preece's block. Russell's S. N. block. Single-needle block.	Electric locking employed between certain electric and mechanical signals.

Block signalling having now, under the Regulation of Railways Act, 1889, become obligatory on all British railways, much of that interest which hitherto attached to the various types of instruments no longer exists; still, a record of the principal systems which have been adopted is desirable, as indicative of the means under which the traffic of our larger railways is worked.

THE THREE-WIRE SINGLE-NEEDLE BLOCK.

As will be seen from the data previously rendered of the systems of block signalling apparatus in use the three-wire single-needle system is very largely employed.

The fact that this type of instrument, as now used, is capable of rendering three distinct and important indications, as shown in Fig. 38, gives it certain advantages. Figs. 39 and 40 illustrate the complete instrument as employed for the protection of trains moving in one direction. Its construction is too well known to need any explanation. The movement of the needle is the result of the influence of the passing current upon the induced needle which forms part of the spindle represented in Fig. 29, p. 61; but the stop pins, which limit the movement of the needle, are arranged so as to afford the latter greater scope, and so render the indications clear. In a block instrument where the needle is required to afford but few indications, and those generally of a semi-permanent character, rapidity of movement is not, as in an instrument employed for telegraphic message work, of moment. Hence the reason for the enlarged radial movement.

FIG. 38.

The movement of the needle to the right or to the left, is obtained by moving the handle in front of the instrument in a corresponding direction. If required to be maintained in either of these positions, the handle has to be what is termed "pegged" over, i.e. fixed in that position. This has for years been accomplished by means of a steel peg, so arranged that it shall engage with, or ride upon, either side of a double inclined plane, according to the position in which the handle is required to be retained. Reference to Fig. 41 will make this

FIG. 39.

clear. The shaft A of the instrument handle B is bored vertically in a line with the handle. On the handle being held over, the pin C is inserted as shown, and, engaging with one side of the double inclined plane of the block D, retains the handle in the position shown, or the reverse—indicated by the dotted line.

The objection to this arrangement is that the pin, becoming worn, does not secure that exactitude of position necessary to ensure good contact with the circuit springs actuated by the movement of the handle; occasionally the circuit becomes broken, and the signal required to be rendered fails. Again, in course of time the point of the pin becomes excessively smooth, and with the vibration of the signal box is liable to slide up the inclined surface, producing a similar result.

FIG. 40.

The arrangement is altogether crude, and it is somewhat marvellous that it should have continued in use for so long a time. An improved method is that adopted by the author, and shown in Fig. 42. Instead of passing a pin through the shaft of the handle, a short pin A is permanently fixed in its lower side, and a sliding bolt C carried in a frame D fixed to the flange I of the socket of the handle, maintains it rigidly in whichever position it is required to be placed. The pin A engages with the bolt C in the same manner as the old form of pin does with the double inclined plane. The handle is held over to either side as required; the forefinger of the operator presses forward the bolt by means of the projection F, and the handle is securely locked. To release it the signalman has merely to hold the handle so as to ease its grip of the bolt, and, applying his finger to the inner side of the part F, withdraw the bolt, when the handle will resume its vertical position. The arrangement has been appropriately named the *Trigger lock*. The whole of the block instruments in use on the Midland system, and it is believed a large number of those in use on other lines, now employ this improved system of operating this class of block. Apart from the greater security it affords in the rendering and retention of signals, it will be observed that, whereas with the old pin and chain arrangement the signalman is called upon to employ both hands in order to operate it, the *trigger lock* is, with ease, dealt with by the one hand, leaving the other free to deal with any neighbouring signal lever which may call for ready manipulation.

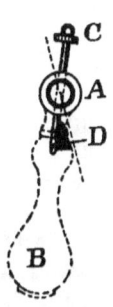

FIG. 41.

Two such instruments are required for the end of each section, as stated, one for the up, the other for the down road. The governing instrument is that fixed at the end of the section which is being approached by the train; thus in diagram

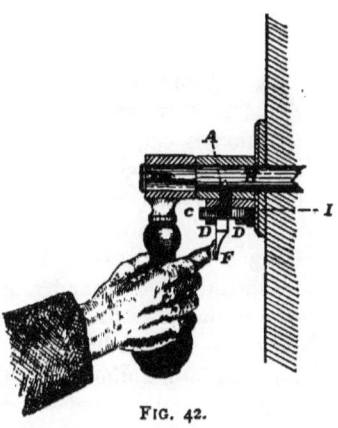

FIG. 42.

Fig. 43, the instrument x is the governing instrument for the road A to B, and y that for the road B to A. If A has a train ready to proceed to B, the indication on his instrument being *line blocked*, A has to ask B by means of the bell for permission to start the train. If the line is clear and B is ready to receive the train, he will peg, or lock, his (x) instrument handle over to the *line clear* position, when A will start the train, and on its entering the section its departure will be signalled to B. B will then unpeg, or unlock, the handle of x and reverse its position, thereby indicating *train on line*, and retain it so until the train arrives at B and has been signalled forward into the next section in advance, or cleared the signals which govern trains at B coming from A to B. In like manner the instrument y will govern trains in the opposite direction.

Where this form of instrument is used for single line working, although desirable, for the reasons which will hereafter be explained, that three wires and a similar disposition of the instruments should be employed, two wires only, one

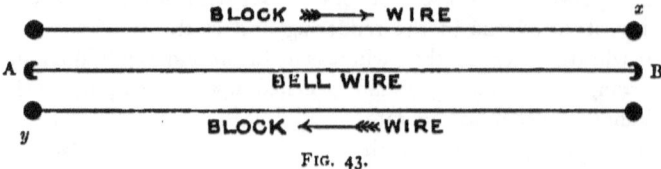

FIG. 43.

for the bell and the other for the block, are at times made use of. In this case the block instrument at either end of the section is employed for both up and down trains; both stations have the power to peg, or lock, over their needle, and consequently that at the other end. This is not desirable. With properly drafted instructions, and careful observance of them by those entrusted with the working of the instruments, there should be no misunderstanding; but it will be seen that if each station were to "peg over" at the same time no signal could pass—the needle would be vertical. This, of course, should not form authority for the passing of a train into the section, but it might lead to confusion and dispute. A man *may* peg his own instrument over for the train which he is about to start. It is better, whether for double or single lines, to provide an independent wire for the government of the trains passing in each direction—one for the up trains and one for the down trains—and for single line working to interlock the two block instruments, so that when permission is being given by one station for a train to approach, it shall not be possible for the other instrument to be operated. Misunderstanding or confusion is thus rendered impossible.

This can easily be effected by inserting a spring contact in the governing—i.e. the x and y—instruments, so that when the handle is turned for either the *line clear* or *train on line* signal, the action shall raise this spring and cause it to break circuit. If, now, this spring with its contact piece is inserted in the line wire of the instrument provided for trains travelling in the *opposite direction*, it will be clear that whenever the governing instrument for the approaching train is operated, that for trains in the opposite direction will be rendered *hors de combat*, so that it would be impossible to signal a train in the direction to *meet* that which had already been signalled.

It is most undesirable that any form of block signalling instrument should be used for any other purpose than that for which it is provided. The single-needle block is capable of being used as a telegraph instrument. Messages may be signalled by it, more slowly it is true, but still in the same manner as by the single-needle message instrument. This is an objection frequently urged. The employment of the instrument for such purposes by the men appointed to work them should be, and invariably is, imperatively forbidden.

The *bell* instrument usually employed with this, as with most forms of block, is that originally designed by Mr. W. H. Preece, supplemented by a *key* for operating it, and which will be found more fully described later on.

SPAGNOLETTI'S BLOCK.

The Spagnoletti block has been fully described in the author's previous work on the 'Application of Electricity to Railway Working.' It remains practically the same. It is a three-wire system, but in the place of the single-needle dial we have a screen with an aperture in the centre, as shown in

Fig. 44. The needle carries a flag inscribed "Train on line" and "Line clear," on either side of the vertical line formed by the needle when at rest. The positive current exposes to view through the aperture referred to, and more clearly represented in Fig. 45, one of these inscriptions, and the negative the other, thus conveying to the signalman a direct instruction. The needle when not conveying one of these signals is dormant, and the flag hangs midway, exhibiting but partially

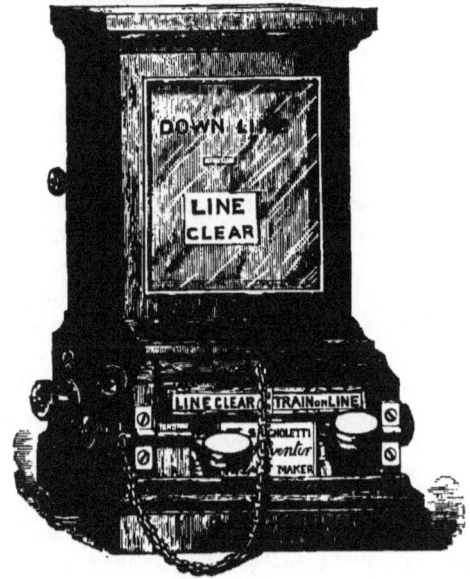

FIG. 44.

the inscription on either side. This is tantamount to "Line blocked," for, as with the single-needle system, the signalman, before allowing a train to enter a section, has to obtain permission from the man at the distant end, and, as will be understood, this has to be conveyed by causing the indication "line clear" to be exhibited.

The method of operating the instrument, although in principle the same as that employed in the single-needle,

to Railway Working.   95

differs to this extent. Two tappers are employed instead of the vertical handle, one, coloured red, for the "train on line," the other, coloured white, for the "line clear" signal. These tappers are so arranged that they cannot both be

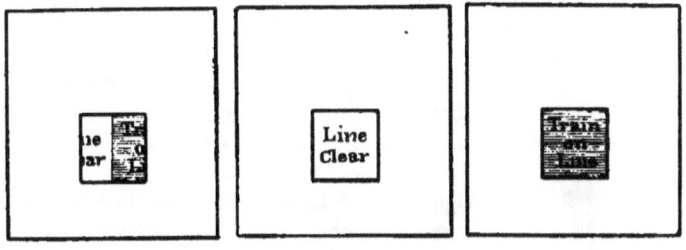

FIG. 45.

pressed down at the same time, so that there can be no confusion of the signal required to be sent.

Any form of single-stroke bell may be employed in conjunction with these instruments.

PREECE'S THREE-WIRE SYSTEM.

Independent of certain electrical advantages, the chief characteristic of this system presides in its assimilation, in form, as well as in the mode of working, to the mechanical signals employed for the guidance of the engine driver. In the early days of block signalling, one of the difficulties which had to be overcome was that attending the instruction of those to whom the working of such apparatus must, of necessity, be entrusted. The interlocking frame was then in its infancy, and men were not so ready as now to grasp the principles of block signalling, or to recognise the many advantages attending either the one or the other. Mr. Preece conceived the idea that if the block-signalling apparatus were so constructed as to be in form and manipulation akin to that employed for the outdoor or line signals, signalmen would readily understand both their purpose and the mode of working them. Hence he constructed as the block signal an instrument, Fig. 46, having a pillar and arm similar to the semaphore, and a switch, Fig. 47, for operating it, similar in its action to the signal-frame lever.

The bell, Fig. 48, employed in connection with these instruments is supplemented by an indicator, the action of which is dependent not only upon the operation of the bell key, but also upon the position of the arm of the block instrument governing the road in question. The arrangement pursued by the inventor will be understood by a perusal of Fig. 49. The position of the lever operating the signal arm determined the direction of the current, whether positive or negative, which the bell key, when depressed by the signalman, should send. According to the nature of this current, so is the action of the indicator of the bell, causing the inscription on the face thereof to read Signal OFF, or Signal ON, at —— whatever may be the name of the signal post.

With the three-wire single-needle block, and with Spagnoletti's three-wire system, we observe that the same description of instrument is employed at either end of the wire provided for the protection of each road. With Preece's system this is not so. The semaphore instrument is that which governs the approaching train. If the arm is lowered, it indicates that the line is clear and that the train may proceed. On the train being signalled forward the arm is raised to the danger position, and there retained till the train has passed out of the section. The semaphore is operated by the switch, Fig. 47. We thus have on this wire, the semaphore at the governing end and the switch at the operating end. Every bell signal sent by the signalling post at which the electric semaphore is placed, records, by the bell fixed at the switch station, the position of the semaphore arm. The signalman, at the latter

FIG. 46.

to Railway Working.

post has thus before him a constant record of the condition of the arm which he manipulates at the distant station. The bell is to him, in fact, an electric repeater.

FIG. 47.

The system of working usually pursued is to clear the section for each train as it passes out of the section. The signals which indicate to the driver whether his road is open or not, are operated strictly in accord with the electrical signals; but it will be clear that, if so desired, the electric, as also the line signals, can be worked in the opposite manner, i.e. by keeping them at danger and having them cleared before a train is allowed to enter the section.

It is worthy of note that this is the only system which records at the operating post the condition of the block signal at the distant signalling station; for although with the single-needle, Spagnoletti's, and in fact all other systems, the sending, or operating, instrument is assumed to work in accord with that at the distant block post, it is by no means impossible for it not to do so. If the wire were to *earth*, the instrument at

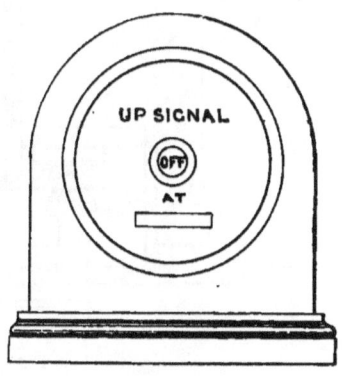

FIG. 48.

H

98   *The Application of Electricity*

either end could be worked independent of, and without in any way operating, that at the distant station. Such action would not, of course, constitute a signal, or meet the requirements for signalling a train. The Code of Instructions under which

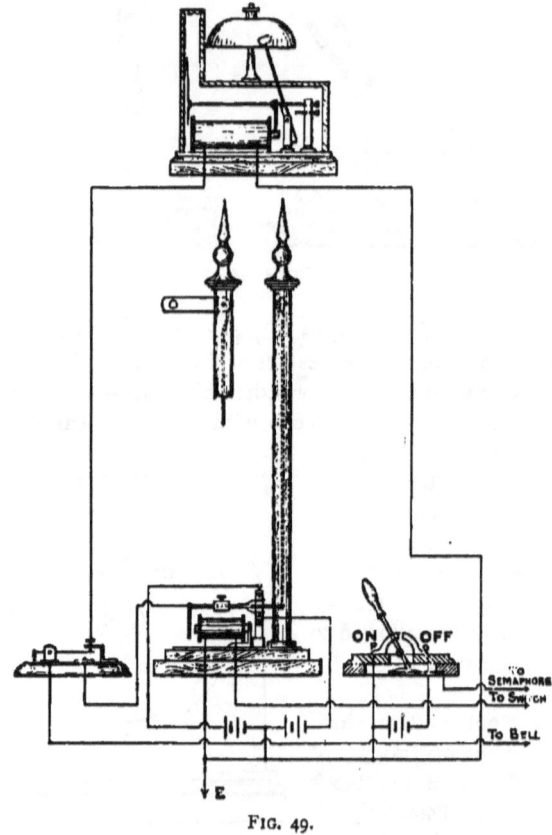

FIG. 49.

the signalling has to be conducted should, naturally, provide checks for any misunderstanding, or irregularity which the form of instrument may be guilty of committing. With Preece's the record, as will have been gathered from the

to Railway Working. 99

previous explanations, and from a study of the connections of the apparatus, is received on another wire from that on which the block instrument is fixed. It is true that if the semaphore wire were to earth, the current sent by the switch would not reach it. The signalman at the semaphore end would notice that his instrument was not operated, and he would call attention to the matter by the bell, and the fact of his doing so would tell the switch station, by the indicator of the bell, that he had not operated the semaphore.

PREECE'S SINGLE-WIRE SYSTEM.

Apart from the fact that it is a single-wire system, and, consequently worked by momentary currents, the chief characteristics are the same as those of the three-wire system just referred to. The switch and bell key are the same ; the semaphore and bell are combined, as shown in Fig. 50.

FIG. 50.

The switch in this case does not, however, directly operate the semaphore arm. It simply directs the current which is to be sent by the bell or signalling key—whether positive or negative. The signalman, should he require to place the arm at the distant station at danger, first throws over the lever of the switch to the ON position, which, say, connects the positive pole of the battery to the signalling key. On pressing down this key, the current passes to the distant station, raises the arm there, and rings the bell as many times as the signalling key is pressed. If the arm has to be lowered, the switch is placed in the OFF position, which reverses the battery current; but in this instance, as will be seen later on, the arm is not operated direct by the current from the distant station. This signal, for its completion, requires the concurrent action of the signalman at each end of the wire.

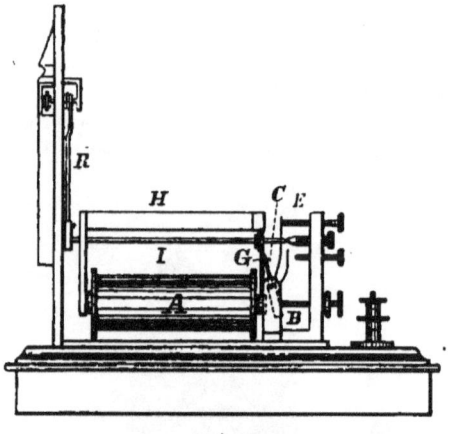

FIG. 51.

In Fig. 51, which represents in side elevation the lower portion of the block instrument, and in Fig. 52, which is an end elevation of the same, we have the internal arrangement of the signalling portion. A is an electro-magnet. To the armature B is attached a locking bar C and a tension spring E. The duty of this armature is to close or open the local bell circuit as required, and by the aid of the locking bar C to lock the indicating portion, which operates the arm O, in the position afforded it by the last current received.

The arm O is operated by the induced magnet G, which

to Railway Working.

is fixed upon the spindle I, and which is connected by a crank lever to the rod R. G, it will be observed, plays between the poles of the electromagnet A A', and obtains its magnetic life from a permanent magnet H, which is connected at its opposite end to the poles of the electro-magnet, imposing upon them a magnetic influence, which is only temporarily overcome by the current from the distant station when passing through the coils. The influence upon the induced magnet G is, normally, that of H, which imparts to it a north polarity at its lower extremity, and which, at the same time, imparts a north polarity to the poles of the electro-magnet on

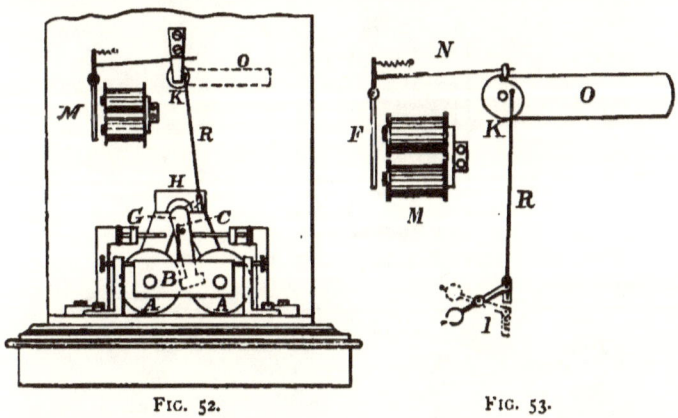

FIG. 52.   FIG. 53.

either side of the S pole of G. G is therefore held by this magnetism to either of the two poles of A, A', against, or in the neighbourhood of which it may be placed. On a current passing through the coils, the poles A, A' will be magnetised in accordance with the direction of that current. One pole will have its magnetism increased, while that of the other will be either reduced or reversed, and G will be influenced thereby to assume the position desired, either that which will raise the arm or prepare it for being lowered.

In Fig. 53 will be noticed a small electro-magnet M. The rod R, which raises the arm, works in a slot, so that it (the crank lever fixed on the spindle I) can be depressed without necessarily lowering the arm. Affixed to the arm is a snail piece K, the projection of which is so arranged that, when in its normal position, the extension N, attached to the armature piece F of the small electro-magnet M, shall, when the arm is raised, engage with it, and retain the arm in the danger position until released by N being lifted away from the snail piece. This is effected only when a current is sent through M.

If we now turn to Fig. 54 we shall readily follow the electrical connections. The indicator coils are in circuit

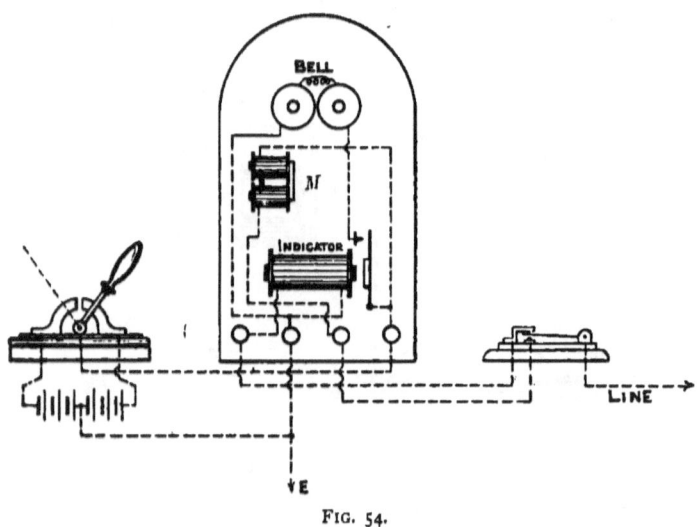

FIG. 54.

with the line wire. The coils M are in circuit with the signalling key or plunger, and the bell coils are arranged on a local circuit actuated by the armature of the indicator coils.

An incoming current passes through the plunger to the semaphore instrument, and through the coils A to earth. The armature B is attracted, the local bell circuit closed, and the bell sounded. If the current is in that direction required to raise the arm to the danger position, the induced needle G is moved to the right, and the arm thereby raised. On the cessation of the current the armature B, under the influence of the tension spring attached to it, falls back, and the locking bar C then interposes to prevent any unauthorised movement of the signal arm.

On a reverse current being received, the armature being attracted, sets G free, when it is carried over to the left. This does not lower the arm, but merely brings the crank to the lower portion of the slot. The arm is held in the danger position by the armature of the coils M.

The receipt of each signal is acknowledged by the depression of the plunger. It will be seen that the coils M are in circuit with the battery and the battery stud of the plunger, and consequently, in sending the acknowledgment, the current passes through the coils M, the armature of M is attracted, and the arm O released. The signal then falls to line clear, or caution, as required. It is only on the acknowledgment being sent that the line clear signal is completed.

TYER'S BLOCK.

The block signalling instruments invented by Tyer, although varying in detail, are in principle essentially the same, consisting of two indicators; one, the upper, being the governing signal for the section to which the instrument applies; the other, the lower, being an index of the last signal sent to the distant station. These indicators may be needles or pointers, or they may be semaphore arms as shown in the accompanying cut, Fig. 55, which represents a complete set of apparatus for the end of one section. The arrangement, as will be observed, embraces a bell or gong, the block instrument is fitted with a bell key or "plunger"; the latter being arranged so that by twisting it round you may obtain a reversal of the current capable of rendering the signal conforming to that presenting itself in the circular space above the plunger.

The indications as represented in the illustration show

that both up and down roads are blocked, or that an up and a down train are travelling between the signal post at which

Fig. 55.

the instruments before the reader are assumed to be fixed, and the adjoining post. The method of signalling would be as follows. The approach of a train is announced on the bell to the post in advance. If the road is clear the signalman will already have lowered the upper arm, which will be an indication that the train may proceed. On the train entering the section, its departure will be announced, and the signalman at the post in advance will then raise the top (red) arm to the danger position, and retain it so until the train reaches him and passes out of the section, when he will lower the arm to the *all clear* position. The system of working is not in all cases the same. In some instances the arm is lowered only on the announcement of the approach of the train—say when it has been received in the preceding section, and the fact of its clearing the section is merely announced to the post in the rear by a bell signal.

The governing post when lowering or raising the block signal arm at the distant station, operates in a corresponding manner the lower, or white, arm on the face of his own instrument; the current employed to operate the distant (red) arm at the same time operating his own (white) arm signal. The miniature semaphore post shown on the face of the instrument may be regarded as a signal post provided with signal blades for governing up and down trains. In the illustration there is, however, but one governing signal, the upper arm; the lower one being but a reflex of the governing signal at the distant post.

Some instruments are provided with two plungers for operating the indicators or arms, one for raising the arm, the other for lowering it; and an independent plunger for the bell. Others have the plunger mounted upon a switch which is turned in one direction for rendering the *blocked* and in the other for rendering the *clear* signals. The manipulating devices are, as has been suggested, numerous, but the principle throughout is the same. The *line blocked* signal is the result of the current in one direction; the *line clear* signal that of a current in the opposite direction. It will be clear the manipulation of these currents can be readily accomplished in various ways, by means of a switch or by means of separate plungers. As in Preece's single-wire system, the direction of the current required to render the signal has to be prepared before it is, by pressing the plunger or signalling key, brought into operation.

These instruments as generally employed are based upon the single-wire system. The movement of the indices, or arms, is produced by the movement of a polarised magnet, which is maintained in the position in which the last current has placed it by the residual magnetism of the cores of the

FIG. 56.

two coils between which the induced magnet plays; these cores being of steel retain the magnetic polarity imparted to them by each passing current.

Although the majority of these instruments are, as stated above, constructed upon the single-wire system, Messrs. Tyer provide for a three-wire system where required. The instrument, Fig. 56, is, it will be seen, very similar to the one-wire form of instrument. The needle of the upper screen A is the block signal governing trains proceeding to the next station, and is worked therefrom. That of the lower screen is operated by the outgoing current of the station at which the instrument is fixed, and forms a reflex of the signal last sent for trains approaching.

B is a switch, operating, mechanically, the indicator seen in the rectangular space above it. This indicator carries three inscriptions conforming to those seen on the dial, viz. "Train on line," "Line closed," and "Line clear." D is the bell key or plunger.

On a train being announced from, say station B, station A, if the line is clear, will turn the switch B to line clear. This will place a current on the wire serving, in this case, the down line, which will deflect A's lower indicator, and B's upper (or block) indicator to "Down line clear." On the departure of the train being announced by B, A will reverse his switch, thereby reversing the direction of the current, carrying his lower needle and B's upper needle over to "Down train on line." And on the train reaching A and passing out of the section, he will turn his switch to the position indicated in the plate when the indication on the dial will be also as there shown.

The plunger D, it will be observed, works within the cylinder which forms a portion of the movable part of the switch. The three wires are apportioned in the usual manner—one to the up and one to the down block, and the third to the bell, which is not shown here.

PRYCE AND FERREIRA'S THREE-WIRE BLOCK.

This instrument, serving an up and a down line, is shown in Fig. 57. It will be observed that it affords three indications, similar, although the nomenclature adopted is not quite the same, to those of the single-needle block. The needles have a greater angular movement than is usual with this form of instrument—the result of the internal arrangements. The divisions on the dial, which indicate the several descriptions of signals, are coloured to accord with the general system of out-door signals :—White, for " Line closed " ; green, for " Line clear " ; red, for " Train on line." The knob which operates the commutator has a segment painted to correspond with the dial. The outgoing current passes through the coils actuating the lower needle, which, together with the positions of the commutator handle, forms an index of the last

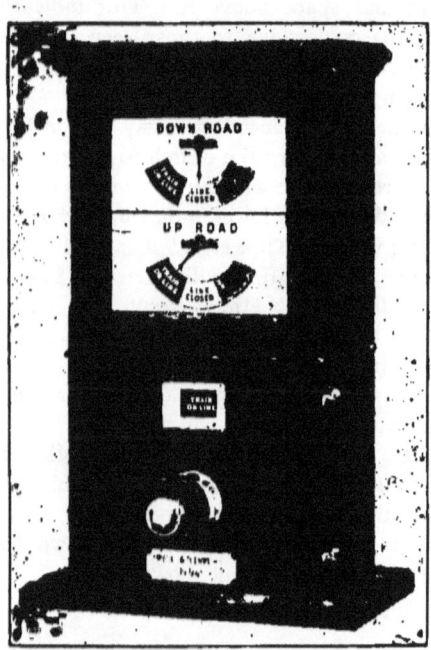

FIG. 57.

signal sent. The ringing key or plunger is seen in advance of the knob which actuates the commutator. As all signals pass through the bell, no movement of the needles can take place without ringing the bell at the distant station.

The internal construction is shown in Fig. 58. A is the dial plate ; B, B' cylindrical armatures, magnetised by the extended pole pieces D, D' of the permanent magnets C, C' ;

E, E' are the coils of the electro-magnets, which by the aid of their prolonged cores F, F' produce, according to the direction of the current, the movement of the needles.

The poles of the permanent magnets attract the sections of the cylindrical armature when uninfluenced by the pole pieces F, and produce the "line closed" signal. The pole

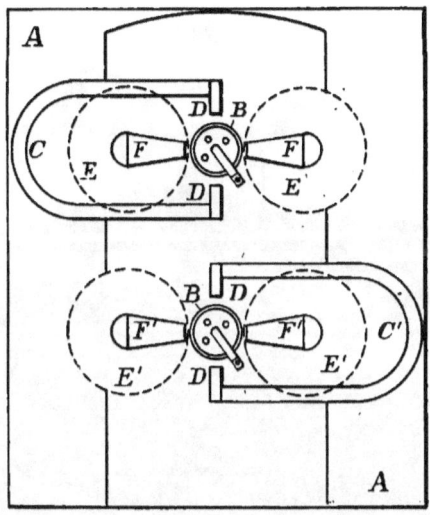

FIG. 58.

pieces F are magnetised according to the direction of the current passing through the coils, and direct the movement of the needle accordingly to "train on line" or "line clear" as may be required. The currents producing these signals are permanent currents for the time being; and so long as the current prevails, so long is the signal maintained.

THE BLOCK BELL.

The instrument represented in Figs. 59 and 60, originally devised by Mr. W. H. Preece, may be accepted as the most suitable bell for block signalling purposes. With a proper adjustment of battery power, tension of the armature spring, and the distance which the hammer is required to travel before striking the bell dome, the best results may be obtained from it. Amply loud enough for any signal box, it is yet capable of rendering the beats with that rapidity and clearness

FIG. 59.

so necessary to successful block working. It is a direct-action bell. The coils are usually wound with No. 26 wire (·018 S.W.G.) to a resistance of 45 ohms. The *key* is retained in its normal position by a spiral spring working within the small cylinder seen in front of the bridge piece, or it may be similarly influenced by a spiral spring in tension at the back

to Railway Working.

of the key lever. The author has adopted the former method as less liable to failure. The almost constant use of the key throws so much stress upon a spring in tension that the bearing parts soon wear through, and, quite independent of the abrasion, not unfrequently break. The spring in compression is not similarly affected.

Fig. 60 shows the bell with its case removed, and Fig. 61 the spring arrangement previously referred to. *a* is a stud

FIG. 60.

fixed to the key lever, so arranged that it shall work within a cylinder *b*, which contains a spiral spring of sufficient power to ensure contact with the back stud. At *d* a flexible connection is provided to ensure good electrical continuity between the bridge piece and the lever of the key, and so prevent the creation of resistance in the line connection.

Where several bells are required for use in the same signal box, each bell should possess a sufficiently distinctive sound

112   *The Application of Electricity*

to enable the signalman to distinguish the one from the other. Where this cannot be satisfactorily accomplished an *indicator*

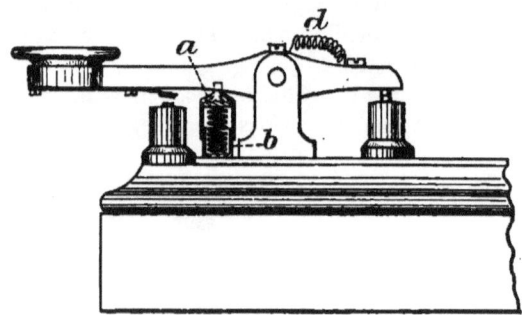

Fig. 61.

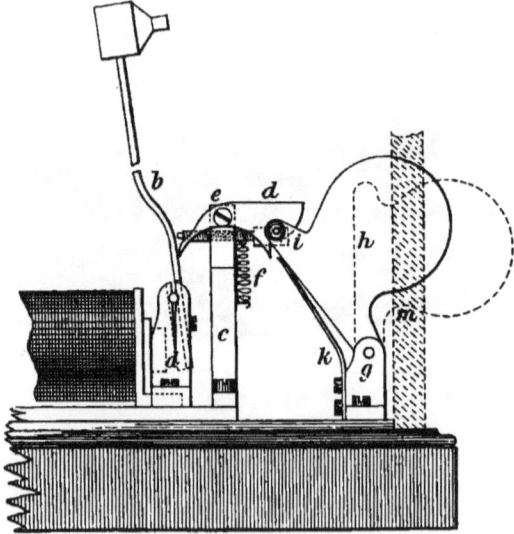

Fig. 62.

bell may be employed. The indicator may be a disc, so arranged that on the armature being attracted it shall discharge a disc, which, falling outwards from the front of the case becomes readily visible to the signalman. This form of indicator is shown in Fig. 62. To the armature a is attached the spindle b of the bell hammer; and mounted on the cock c, centered at e, is a tripping piece d, which at the b end is bent round at right angles, so that any forward movement of b shall press against it; d has a shoulder piece which limits its movement in the direction of the spiral spring f, which is provided in order to retain d in the position indicated in the illustration; h is the indicator, painted red, pivoted at g and carrying at i a small pin required to engage with d in the semicircular slot in which it is shown; k is a flat spring pressing against the back edge of h, the duty required of which is to drive the latter forward immediately it is released at i; m represents the front of the bell case.

On a current passing through the coils, a will be attracted, b will press against the tail end of d with sufficient force to depress it, and so release the pin i. h is now, under the influence of the spring k, thrust forward into the dotted position. The indicator is thrown back into its position of rest by the signalman's finger when next operating the bell key, or the depression of the key itself may be made to accomplish the same duty.

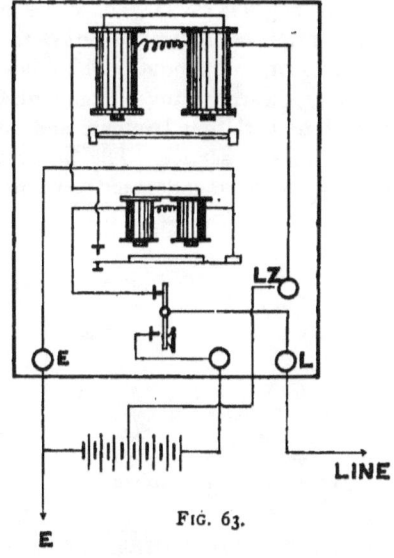

FIG. 63.

Where block sections are switched through at night, so as to form one long section, it is of advantage to employ *relay bells*—bells, the relay operating which is in the line circuit, as indicated in Fig. 63.

The electro-mechanical bell, used to so large an extent some years since, has now become obsolete. The direct-action bell, such as that previously referred to, is that which alone can meet the requirements of a busy and economically managed line of railway.

The bell is the directing voice of all block signalling systems. It would be as reasonable to depute a dumb man to direct the operations of an army corps as to seek to establish a block service without the aid of a bell. The necessity for a *bell code* to be used in common by all railway companies has been a long-felt want. The codes made use of, although in many instances similar, all varied to some extent; they were not uniform throughout. The following code, known as the Railway Clearing House Standard Bell Code, has now been agreed to, and, to all intents and purposes, adopted by the entire railway service. The importance of this step where railway systems under independent management converge will be self-evident.

BELL SIGNALS.

	Beats on Bell.	How to be given.
Call attention	1	1.
Train entering Section	2	2 consecutively.
Is Line Clear for Express Passenger Train, or Breakdown Van Train going to clear the Line, or Light Engine going to assist disabled Train?	4	4 consecutively.
Is Line Clear for Ordinary Passenger Train or Breakdown Van Train not going to clear the Line?	4	3 *pause* 1.
Is Line Clear for Branch Passenger Train?	4	1 *pause* 3.
Is Line Clear for Fish, Meat, Fruit, Horse, Cattle, or Perishable Train composed of Coaching Stock	5	5 consecutively.
Is Line Clear for Empty Coaching Stock Train?	5	2 *pause* 2 *pause* 1.
Is Line Clear for Fish, Meat or Fruit Train composed of Goods Stock, or Express Cattle or Express Goods Train?	5	1 *pause* 4.

to Railway Working.   115

BELL SIGNALS—*continued.*

	Beats on Bell.	How to be given
Is Line Clear for Ordinary Goods or Mineral Train stopping at intermediate Stations?	3	3 consecutively.
Is Line Clear for Branch Goods Trains?	3	1 *pause* 2.
Is Line Clear for Through Goods, Mineral or Ballast Train?	5	4 *pause* 1.
Is Line Clear for Light Engine, or Light Engines coupled together, or Engine and Break?	5	2 *pause* 3.
Is Line Clear for Ballast Train requiring to stop in Section, GOODS TRAIN CALLING AT INTERMEDIATE SIDING IN SECTION, or Platelayers' Lorry requiring to pass through tunnel?	5	1 *pause* 2 *pause* 2.
Train out of Section, or Obstruction removed	3	2 *pause* 1.
Bank Engine in rear of Train	4	2 *pause* 2.
Obstruction Danger	6	6 consecutively.
Blocking Back	6 {	*Inside* Home Signal—2 *pause* 4. *Outside* Home Signal—3 *pause* 3.
Stop and Examine Train	7	7 consecutively.
Cancelling "Is Line Clear," Signal or "Train entering Section" Signal	8	3 *pause* 5.
Train passed without Tail Lamp	9 {	9 consecutively to BOX IN ADVANCE. 4 *pause* 5 to BOX IN REAR.
Train divided	10	5 *pause* 5.
Shunt Train for following Train to pass	11	1 *pause* 5 *pause* 5.
Vehicles running away on wrong Line	12	2 *pause* 5 *pause* 5.
Section Clear, but Station or Junction blocked	*13	3 *pause* 5 *pause* 5.
Vehicles running away on right Line	14	4 *pause* 5 *pause* 5.
Opening of Signal Box	15	5 *pause* 5 *pause* 5.
Testing Block Indicator and Bells	16	16 consecutively.
Closing of Signal Box	17	7 *pause* 5 *pause* 5.
Time Signal	18	8 *pause* 5 *pause* 5.
Lampman or Fog Signalman required	19	9 *pause* 5 *pause* 5.
Testing Slotted Signals	20	5 *pause* 5 *pause* 5 *pause* 5.

* Employed only subject to stoppage of train and verbal warning of driver, in addition to exhibition of signal to proceed at caution, and at such points as are specially authorised.

The adoption of this code does not preclude the interposition of other codes where found necessary for local purposes. Such additions are in fact made, being distinguished from the general code by the employment, for them, of a distinctive form of type. Nor, so long as the number of beats rendered corresponds with the full number shown, does it matter how they are rendered, whether with pauses between, or not. Some companies prefer to rely upon the bell code entirely for the description of train, whilst others by various ticks of the needle, where the single needle block is employed, render many other signals than are enumerated in the bell code. These bell signals are acknowledged by repetition by the station to which they are rendered, as soon as that station is in a position to accept the signal indicated.

JUNCTION WORKING.

The following are the regulations now generally adopted for working traffic over junctions :—

When permission has been given by B for a train to approach from C, no train must be allowed to leave D until that from C has been brought to a stand at the home signal, or has passed through the junction for a distance of a quarter of a mile, or until the "train out of section" signal for the previous train has been received from the next signal box ahead if within that distance; nor in such a case must a train be allowed to leave A for D unless the junction facing points at B are set for C, and the line towards C is clear for a distance of a quarter of a mile beyond the junction points, or until the "train out of section" signal for the previous train has been received from the next signal box ahead if within that distance.

When permission has been given by B for a train to approach from D, no train must be allowed to leave C until that from D has been brought to a stand at the home signal, or has passed through the junction for a distance of a quarter of a mile, or until the "train out of section" signal for the previous train has been received from the next signal box ahead if within that distance.

When permission has been given by B for a train to approach from A for D, no train must be allowed to leave C until that from A has been brought to a stand at the home signal, or has passed clear of the junction, or the junction facing points have been set for C, and the line towards C is clear for a distance of a quarter of a mile beyond the junction points, or until the "train out of section" signal for the previous train has been received from the next signal box ahead, if within that distance.

When a train has been sent to the starting signal, and the rear of the train is well clear of the junction, permission for a

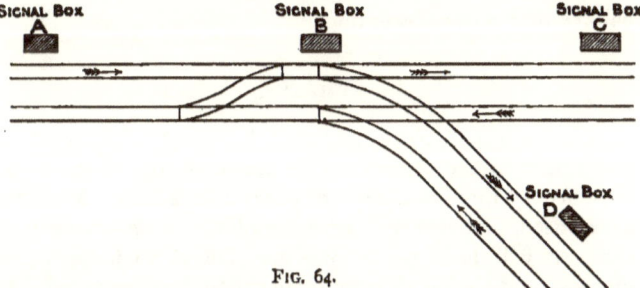

FIG. 64.

following train to approach may be given by the signalman to the signal box in the rear if the points are set for the following train to pass on to another line and that line is clear.

Most block instruments admit of electric interlocking, the one with the other, so as to prevent the "line clear" signal being rendered at the same time for trains approaching on conflicting roads. Where such can be carried out it would add materially to the safe working of such junctions, and would entail but slight additional expense. In the junction represented in the illustration, such a provision would entail the establishment of an independent block circuit for trains travelling from A to B, inasmuch as those for D as well as those for C pass over the same section of line between A and B, and it is necessary that a train from C for A shall not delay a train from A which has to proceed to C, whereas if the train from A is for D, it is necessary it should not pass B until that from C is clear of the junction points. Usually this interlocking of the electrical instruments may be accomplished by arranging the wires so that the operation of the commutator of one instrument for the production at the distant station of the "line clear" signal shall interrupt the line wire, or the battery wire of the instrument capable of giving the same signal on a conflicting road. It is true that under the Railway Regulation Act, junction signals and points are mechanically locked, but if it is desirable junctions should be worked as is indicated by the foregoing instructions, and of this there can be no doubt, then it is desirable the electrical block should be interlocked so far as possible to enforce that mode of working.

REMARKS.

Although the block system, when once established, is paramount in the control of the traffic, it should never be regarded by those in charge of trains as infallible. Mistakes have arisen, and will still arise. Failures of an unexpected character will develop. Signals may fail at an inopportune moment, and they may be misunderstood or disregarded. It is therefore desirable, when a train is brought to a stand from some exceptional cause, that the prescribed steps for its protection should in every instance be most strictly observed.

A "block" signal, when once rendered, however irregular it may be, should never be disregarded. When maintained longer than usual, it should be, if possible, more than ever fully enforced. It is far better to detain a train than to run any risk. In all such cases the signalman should be governed by the printed instructions under which the block is to be worked.

In like manner, should an unusual time elapse before a train signalled into a section arrives at the post in advance, the signalman there should have no hesitation in putting his mechanical signals for trains approaching in the opposite direction, at danger, in order that he may stop and caution the driver to proceed with care in case the delayed train may have fouled the opposite road.

The mode of fixing the "block" instruments over the signalman's lever-frame, advocated in the author's previous work, is now almost invariably followed by all railway companies. Almost every form of block instrument—with the exception of the tablet or electric staff—is capable of this arrangement. It is important that the instruments should be always before the signalman; and the saving of time to the signalman, as well as the convenience attending the manipulation of the apparatus when so arranged, are all points of considerable importance in the economy of the signal box.

Where "repeaters" are employed they should be placed in such a position as will readily identify them with the signal levers employed to operate the signals to which the repeaters refer. They may be placed, each one on the front of the instrument shelf, over, or immediately behind, the signal lever to which they apply.

Light indicators occupy more room, and as they are, or should be, provided with a bell to ring continuously when the light is unsatisfactory, they may be arranged on a shelf over the block instruments, or other convenient place. When provided with a continuously ringing bell, it is not so necessary that they should be immediately under the eye of the signalman, as the bell will, when necessary, attract his attention.

It is undesirable block instruments should be tested by the lineman or others, when a train is being signalled by them, and in no instance should the batteries or wires be disturbed without the knowledge and approval of the signalman. In a similar manner the mechanical signals should not be moved when a train is approaching, or under their protection. The lineman should, when necessary to operate a signal arm, or other part of the signalling gear, do so only under the sanction of the signalman, who will, of course, take steps to safeguard the line.

In the author's previous work certain principles which

should govern the selection of a block system are laid down. In the majority of instances these have received recognition. Those systems which entail the employment of a constant current for the block or clear signals are being more generally employed, to the displacement of the momentary or transient current systems. Where possible, as in the single needle, and in Tyer's, and Pryce and Ferreira's, the normal or gravity position of the signal now constitutes the "line blocked" signal. The objection to the employment of any form of instrument, capable of being employed for conversational purposes, for block signalling still remains good, and that even in the face of the enormous number of single-needle block instruments in use. At the same time it must be admitted that the ready means now so generally afforded signal boxes for conversation by means of the telephone, does, where this provision is made, reduce to some extent the force of the objection. The desirability of affording all adjoining signal boxes a means of *vivâ-voce* communication, either by independent telephone circuits or by combining the telephone with the block bell, has been advocated elsewhere. It will be found in all instances to be most material to the ready disposal of traffic, and to amply repay the cost of its establishment.

The signals rendered by momentary or transitory currents are capable of being negatived, or reversed by foreign currents from neighbouring wires with which they may make contact, or by lightning; and even signals rendered by permanent currents are liable to be erroneously rendered by the interposition of foreign currents, due to contact with wires carrying currents of the required polarity. For this reason it would appear to be most desirable that every "Line clear" signal should be confirmed, preferably by a bell signal consisting of more than one stroke on the bell. If, by the aid of a foreign current, the "Train on line" signal is produced, the greatest harm effected is probably the stoppage of traffic. That is bad enough on a busy line, but it is not attended with that danger which, with lines worked on the Affirmative system, might attend the formation of the "Line clear" signal when the signalman desired to maintain, and from not having himself rendered the line clear signal, believed the section for trains approaching him was at "Line blocked."

To realise this: assume A has a train which he desires to send to B. B has A's signal at "Line blocked," and is not in a position to receive the train from A, consequently he does not acknowledge the inquiry *is line clear, &c.* But a foreign current at that moment enters the block wire governing trains from A to B, of such a nature as to render the "Line clear" signal. A might possibly, under such circumstances, send on the train. Where the bell signal "Is line clear" is repeated prior to the "line clear" signal being rendered, there is in that action a confirmation on the bell, but it is not so clear or so definite as would be a "line clear" bell signal *accompanying* the block indication of that signal. It will be observed that in the Clearing House code no such signal is provided for. It has apparently been considered that the repetition of the bell signal "Is line clear," or an acknowledgment on the block needle, sufficiently meets the purpose. The point is one which will bear consideration.

It is, of course, clear that the presence of such a foreign current, producing a line clear signal when none was intended, should be noticed by the signalman at each end, and steps taken to counteract it. No doubt if noticed in time this would be done. The main question is, would the confirmation of the *line clear* signal by a bell signal reduce the possibility of error? If so, it is evidently desirable.

All signals should be acknowledged if possible by repetition, or, failing this, in such a manner as shall indicate that they are unmistakably understood. If a signal is not at once acknowledged it leaves the signalman in doubt whether the signal made by him to the adjoining post has been received or not. Further, its non-acknowledgment leads to irritating repetitions. It is far better to acknowledge a signal, even if the train cannot be accepted, than to leave the man in doubt whether the road is blocked, or the communication deranged.

CHAPTER VI.

SINGLE-LINE WORKING.

ORIGINALLY single lines of railway were worked under regulations which appointed certain fixed points at which trains were timed to cross one another—that is an up train met a down train, and each went on its way. When the traffic became very much out of gear—when one train fell greatly out of time—these crossing points had to be replaced by others, and this had to be effected by telegraph messages indited by the Superintendent of the Line, and addressed to the Station-masters at the several stations affected, in the manner indicated in the author's previous work.

This system has now, under the Railway Regulation Act, 1889, been superseded by the compulsory use of the block.

FIG. 65.

With the block must be incorporated the *staff system*. On single lines worked by more than one engine, the one is never used without the other. The *staff*, Fig. 65, is an instrument resembling, somewhat, a policeman's baton. It is provided with projecting rings or studs, at various points; and in form, may be round, square or three sided; the idea being that the staff applicable to any one section of line should by its shape or other peculiarities, as well as by the inscription upon it, be readily recognisable as applicable to that section; that it should fit the receptacle provided for it, and that it should not be possible for that applicable to a neighbouring section of the line to do so. As will be gathered from the foregoing, each staff has engraved upon it the section of line to which it applies.

It is not necessary that the block sections should correspond with the staff sections. The staff is the agent controlling the *crossing points*.

Tickets are employed in conjunction with the train staff. These tickets are of various shapes, corresponding with that of the section of the staff for the section to which they apply. Thus they are round, rectangular, triangular, &c. Again, they are of various coloured cardboard, and have printed upon them in bold type the section to which they apply. The purpose of the adoption of different coloured paper for adjoining sections is, of course, the ready recognition, by those acquainted with the colour applicable to their section, of the proper ticket.

Each ticket has printed upon it instructions similar to that shown in Fig. 66, which is a facsimile of that in use for a section of line.

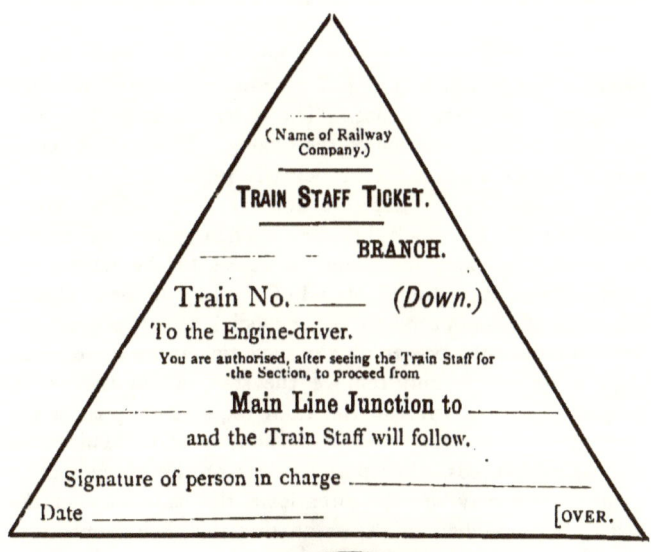

124   *The Application of Electricity*

The reverse is inscribed as under :—

> This Ticket must be given up by the Engine-driver, immediately on arrival, to the person in charge of the Staff Working at the place to which he is authorised to proceed, to be dealt with as the latter may be instructed by the Superintendent of the Line.

These tickets are kept in a box, Fig. 67, which has an orifice through which the staff may be passed, and retained in safety when not in use. The chief object of the box, however, is the protection of the tickets. The staff is necessary to obtain a ticket. If a ticket is required, the staff has to be inserted in the aperture appointed for it, and turned round in order to unlock the box. When turned sufficiently to unlock the box, and admit of access to the tickets, the staff becomes locked until the lid of the box is again closed. Thus the station-master or other official has to possess the staff to unlock the box to obtain such tickets as he requires, then to relock it and remove the staff, fill in the ticket, present it to the driver of the train required to precede the staff, and at the same time show him the staff. This done, the train proceeds. When it is clear of the block section another train may be sent after it in the same manner, the last train proceeding in the same direction being entrusted with the staff. The tickets are carried through to the crossing point, and there collected and sent to the Superintendent of the Line or other appointed officer, to be examined and checked with other documents relating to the progress of the traffic on the line to which they apply. The officer issuing the tickets is responsible that not more than the requisite number of tickets is extracted from the ticket box. Until the staff reaches the crossing point no train is allowed to leave there in the direction from which trains worked under the authority of the staff tickets are arriving.

The advantage of intermediate block sections will, where trains are required to follow one another—one, or more, under

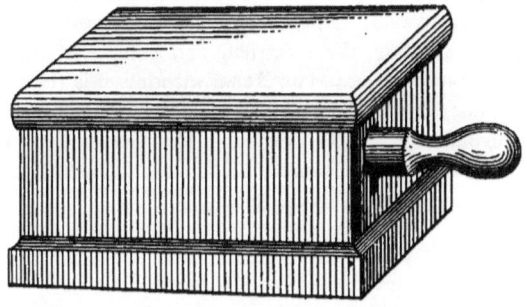

FIG. 67.

a ticket, and the last with the staff—be obvious. It ensures the separation of the trains by a given space.

Safety is, as far as possible, assured by this mode of working, but its inconvenience is occasionally great. The crossing points are fixed. There is no deviation therefrom. If one train is late the other must await its arrival. If the engine of a train breaks down, the staff has to be sent on by mounted or foot messenger, in order to admit of another engine coming to its aid. If a special has to be run, and time has failed to admit of arrangements being made in anticipation, it must await the arrival of the staff.

It was to meet these difficulties that Mr. Tyer introduced his tablet block system, followed later on by the production of the "electrical train staff," by Messrs. Webb and Thompson of the London and North Western Railway. Each system is the product of combined electrical and mechanical action. The tablet—or the electric staff, as the case may be—takes the place of our old friend the train staff, and the principle upon which it is worked is, that the requisite voucher under which the train is to travel—the staff or the tablet—shall, when all is in its normal condition, be obtainable at either end of the section, *but that one voucher only shall be in use at the same time.* The basis of the two systems is practically the same. Given everything in its normal condition, the necessary voucher may be obtained at either end of the section by a current sent from the station in the direction in which the train has to travel. The withdrawal of the voucher, mechanically, so arranges the electrical connections, that no further voucher can be released until that withdrawn is restored either to the instrument from which it has been taken, or to that at the distant end of the section. The electrical arrangement differs in detail, but in principle the result, although obtained by varying methods, is the same.

TYER'S TABLET BLOCK.

This instrument has gone through several stages. Its latest and most complete form, " No. 6," is shown in Fig. 68. It is constructed in two portions, viz. an indicating and a receiving portion, or as Mr. Tyer describes it, it forms an "absolute block" signalling instrument, combined with a tablet apparatus.

The indicating or "block" portion has two discs, upon which are inscribed the words IN and OUT, either of which, according to the conditions under which the instrument is being worked, are caused to appear in the circular openings in the dial plate of the instrument, causing the inscription to read "tablet in" or "tablet out," for the up or the down train as the case may be.

The upper disc is operated electrically, the lower one

mechanically by means of the switch A, seen in the front of the instrument. This switch A operates mechanically a commutator arranged in the interior of the instrument in such

FIG. 68.

a manner that it may be actuated mechanically by A, or electrically by a current from the distant station.

Above the two discs is a current indicator C, the purpose for which will be gathered later on.

The receiving, or the tablet apparatus portion consists of a cylinder for the reception of a series of tablets, together with the necessary electrical and mechanical parts for securing the efficient action of the instrument. B is the *tablet slide*, by means of which the tablets are obtained from the cylinder, and by which they are also restored to the cylinder when done with.

The ringing key, or plunger, is arranged in front of, and operated by, a rod passing freely through the switch A. To the left of the tablet case is a "switch" lever E, which, on the operator drawing it towards him, lifts the tablets in the cylinder, together with the detent, or locking bolt, by which the tablet slide is held in position, so as to admit of the slide being *withdrawn empty* when required.

A bell or a gong is provided with each instrument.

Now, if we follow the working of a pair of these instruments, we shall gain some insight into not only the mode of working, but the principle itself. The upper indicator, it will be remembered, is worked by the current from the distant station. It is, in point of fact, operated by a strong electromagnetic relay, the tongue of which is free to move to the contact provided on either side of it, but the normal position of which is midway between these two contacts. The contact to the left completes a local circuit through a powerful pair of coils, the armature of which operates a detent which requires a sustained current to lift it, and thus liberate the lock on the switch A controlling the slide. The contact to the right completes a local circuit through an equally powerful pair of coils, the duty of which is to control the movement of a commutator governing the direction of all outgoing currents. It will thus be seen that when a continuous current of a certain polarity is passed through the relay coils the lock which governs the slide is released, and when an opposite current is sent, the commutator is operated. And it will have been remarked that to lift the lock of the tablet slide, it is necessary the current should be *continuous*. This is effected by the signalman at the distant station prolonging the pressure on his ringing key. The local coils are so constructed as to require a somewhat prolonged current to energise them, and, in order that they shall not be actuated without the aid of a purposely continuous current, that is, by the ordinary ringing currents, the circuit is formed through a pair of vibrating springs which are set in motion by each stroke of the bell hammer, and only when this hammer, together with the tongue of the relay, is held in a fixed position sufficiently long to enable the springs to become quiescent, is the completion of the circuit sufficiently prolonged to allow of the action of the local locking coils. It may perhaps be thought that herein is a loophole, which may admit of a tablet being extracted when such is not intended. When the instruments are in their normal condition, all tablets being in, the very first ring from the station of whom the demand for permission to withdraw a tablet is made, being in the same direction as that which is prolonged in order to admit of the extraction of the tablet, would, if sufficiently continuous, be equally serviceable for the purpose; but, supposing the circumstances were such as to enable the signalman to do so, no good would arise, and no harm would be done, for the withdrawal of the slide with the tablet in it would at once so alter the arrangement of the instrument, that no other tablet could be obtained, the man would simply have extracted the tablet before he should have done so, and in doing so would have transgressed the signalling regulations.

Let A and B be the terminal stations of a block section, and assume that B requires to send a train to A. B will ring A to that effect, and A, if prepared to admit the train, will, in responding, prolong the last of the series of bell signals, indicating to B that he may withdraw a tablet. This prolonged current energises the local locking coils for the slide and allows B to turn his switch A, so as to cause the lower disc to read "tablet out." This action on the part of B mechanically releases the lock on the tablet slide, and at the same time releases a lever which eventually falls into a groove and prevents the complete return of the slide. B now withdraws the slide and removes from it the tablet which is, in due course, handed to the driver. B is able to push the slide partially home if he desires, but not wholly so, as previously explained. In turning his switch to "tablet out," B reversed the connections of his commutator, and whereas, on previously pressing his bell plunger D, he was able to send a positive current to A, he is now only able to send a negative current; moreover, on turning his switch it became locked. It is now impossible to obtain a further tablet at either end, because the current under operation by B is of the wrong direction to admit of one being withdrawn at A, and B is unable to return his slide to obtain one at that station.

B has now to advise A that he has obtained a tablet, and the current which he employs to do this being of the opposite polarity to that last sent, i.e. before he has turned his commutator, it causes the upper disc in A's instrument to read "tablet out." A is thus acquainted that B has obtained his tablet. The departure signal announces the despatch of the train, and the instruments remain in the position indicated till the arrival of the train at A.

On the arrival of the train at A, the tablet is handed to the signalman, who raises the lever E to the left of the instrument, and, drawing out his slide *empty*, places within it the tablet and pushes the slide home.

The return of the slide *with the tablet in it*, by means of a lever, operated by the tablet as it passes under it, reverses A's commutator. A can now send negative currents only. He acknowledges the receipt of the tablet prolonging the last bell stroke to enable B to reset his commutator to "tablet in." B, on receiving the signal indicating the reception of the tablet at A and its deposit in the instrument there, turns his commutator accordingly. This action on the part of B admits of his restoring his empty slide to its normal position, where it again becomes locked; and it will of course be understood that B's commutator, on being returned to "tablet in," has also become locked.

B is now in a position to communicate positive currents to A, and in acknowledging A's signal, he, by this current, electrically reverses A's upper disc (causing the inscription to read "tablet in") and the commutator in the interior of A's instrument, which, it will be remembered, had been reversed by the passage of the slide with the tablet in it. A is now in a position to communicate positive currents to B. Both instruments are restored to their normal condition, and a tablet may be obtained from either.

We have gone through the process of signalling a train through a section. Suppose, however, that the train did not require to pass to station A, but to return to B. It is necessary the tablet should, in that case, be restored to the instrument at B from which it was taken, and that in doing so the instruments should resume their normal condition. This is effected on the return of the train, by the signalman placing the tablet in the slide and pushing it home. The slide, it will be remembered, could not be *wholly* returned home. It could be retained fully, or partially open—in any case sufficient to enable the signalman to place a tablet within it. This is done, and now on pushing forward the slide with the tablet in it, the presence of the tablet actuates a lever which removes the lock which hitherto prevented the complete return of the slide, before the slide reaches it, and at the same time also releases the lock on the switch A, so that the signalman is now at liberty to turn it to " tablet in." This he does, and in signalling to A that he has received back and restored the tablet to his instrument, he reverses the upper disc in A's instrument, causing it to read " tablet in." A thus learns the tablet has been returned at B, and in acknowledging B's signal completes the local circuit of B's instrument which reverses his commutator, when the normal condition is re-established—each instrument being again in a position to send positive currents to line.

It will be observed that the instrument, the action of which has just been described, ensures by means of the indicators above the switch, a record of the condition of the tablet at each end of the section. Such a record is undoubtedly of great advantage. The signalman is inspired with confidence, he sees what is passing at the other end of the section. It is to him a "block" instrument. At the same time it is not absolutely necessary for the safe working of the traffic. The checks on the issue of the tablet can be maintained without them, and in point of fact an instrument known as "No. 5," Fig. 69, has been produced to meet such demands. Its only indicator is a galvanometer needle I. The tablets are inserted at C, where they fall edgeways into a number of slots arranged in a disc which is rotated by the knob K. The switch S

FIG. 69.

must then be depressed, and the knob K turned from right to left. This will bring a tablet under the opening on the top of the box portion of the instrument, and on raising the cap the tablet can be extracted. The departure signal is then given and acknowledged. On the arrival of the train the cap is raised, the tablet inserted, the cap shut down, the knob turned from left to right, and the "arrival signal" sent. Whenever a tablet remains in the slot immediately under the cap, the battery circuit is disconnected, and no signals can be passed to the distant station; but as soon as the tablet has been taken out of the machine the battery circuit is restored. And whenever a tablet is inserted in the machine the battery circuit is broken. Turning the knob from left to right restores it.

It will readily be recognised how all these actions provide means for accomplishing precisely that which is done under the No. 6 type, minus, of course, the records. The two instruments are, when in the normal condition, in a position to admit of a tablet being obtained from either. The power to release a tablet is obtained by the current from the distant station, in conjunction with the pressure of the switch plunger. The knob K is unlocked; it is turned to bring the tablet to position. This reverses the current, and now no tablet can be extracted from the instrument at the distant station. On inserting the tablet at the distant post, and turning the knob of that instrument, the necessary current for readjusting the commutator at the first station is obtained, and on that station acknowledging the same, the commutator in the latter instrument is restored to its normal condition.

In like manner the principle may be applied, and is applied, to the several types in use.

It has been mentioned that, following the principle of the staff and ticket employed with the block system, "tickets"

FIG. 70.

may be used for trains preceding the "tablet." This is occasionally done, with or without the aid of intervening block posts. But it is important to note that where this practice is pursued, the tablet instrument employed should be of that kind which *does not admit of the tablet being returned to the instrument from which it has been obtained.* A train once sent under a "ticket" must be followed by the "tablet," and if it should happen that there is no train to take it, it only remains for it to be sent to the point required by some other means.

The tablets for each class of instrument are made to a template, in order to ensure perfect fit and avoid friction when in the cylinder of the instrument; but their configuration differs, as will be seen by reference to Fig. 70. Three different configurations are considered sufficient. The object is to prevent all possibility of a wrong tablet being inserted in an instrument. The shape of the notches in the rim of the tablet are those which are required to correspond with the instruments to which they apply.

FIG. 71.

Instead of, as hitherto, working road sidings by fixed signals, a small frame for working the siding points, &c., is provided. In this frame a "key" lever is arranged, and in connection with this lever there is a receptacle for a tablet. When the tablet is inserted in this receptacle, and the slide in which it is placed driven home, or the lid which confines it closed, the key lever can be pulled over. The instant the lever moves, the slide, or the case in which the tablet is placed, becomes locked and remains so until the "key" lever is returned to its normal position, when the lock on the tablet is released, and it can be again removed.

The drawing over of the key lever unlocks all other levers in the frame and admits of their free use. They must, however, be restored to their normal position before the key lever can be restored to its normal position.

In this manner, by the aid of the tablet the traffic at any number of wayside sidings may be dealt with. Without the tablet the points affording access to them cannot be moved, and until the points have again been set for the main road the tablet cannot be set free for the train to proceed.

Fig. 71 is a representation of the form of pouch in which

FIG. 72.

the tablet is placed prior to handing it to the driver of a train. In the centre of the pouch is a space sufficiently large to enable the driver to see the inscription on the tablet, in order that he may be assured he has that applicable to the section he is traversing, or about to traverse. The bow is formed of cane, and its purpose is that when held in the necessary position it may be intercepted by the arm of the driver of a passing engine, or by a projection therefrom provided for the purpose. In a like manner, that which has to be delivered up is handed by the driver to the signalman, or is intercepted by a projection provided at or near the signal box.

Provision is made for switching through sections where required. This involves the establishment at the switching station of a "plunger switch," and at the stations at the end of the prolonged section, of a case in which is a rotating disc, Fig. 72. A written or printed authorisation is sent to the closing station, directing him, after the passage of a certain train, to close the switch for such and such a time. On this being done the following procedure has to be adopted. A tablet is asked for and obtained in the ordinary manner. This tablet is not given to the driver, but is inserted in the opening of the case shown in the illustration marked "round." This unlocks the knob C, and so enables the signalman to turn it one step forward. On raising the lid A, a square tablet will be found ready for extraction. This square tablet is the description of tablet for use during the time the section is switched through. It is taken on to the distant station and there inserted in the "square" opening of a similar box, the knob of which on being turned enables a "round" tablet to be obtained. This round tablet is now inserted in the ordinary tablet instrument and the train announced out of section.

WEBB AND THOMPSON'S ELECTRIC-STAFF SYSTEM.

This apparatus consists of a magazine formed of a cast-iron pillar 3 feet 6 inches in height, having a slot A about $1\frac{3}{8}$ inch wide down the centre of the pillar, and capable of holding eighteen shafts, as represented in outside front and side elevation by Fig. 73, and inside front and side elevation by Fig. 74. Where a larger number of staffs is required, a pillar having a double slot, capable of holding twice the number of staffs, is provided. On the top of the column is fixed what is termed the head of the pillar, and in this are arranged the whole of the mechanical and electrical parts of the instrument. On its face are two indicating discs B and C, and underneath the right-hand indicating disc C is fixed a tapper to *Railway Working.* 137

key F. In the centre of the face there is an ordinary galvanometer or current indicator D. The slot in the pillar, which is provided for the reception of the staffs, is continued,

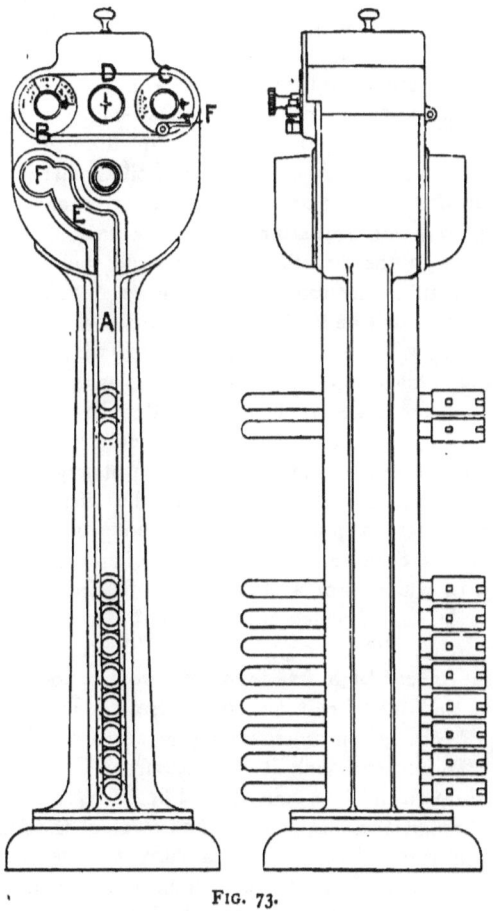

FIG. 73.

as indicated at E, to the left-hand side of the face of the instrument, where it terminates in an enlarged hole F, sufficient for the reception or withdrawal of the staffs one at a time.

In the interior of the instrument are the following electrical and mechanical parts:—Turning on a centre pin G are five metal discs H, H', H'', H''', H'''' secured together. Each disc has four notches at equal distances apart cut out of its periphery, of a sufficient width to admit the shank of a staff. To start with, these notches are so arranged that when a staff has to be inserted or withdrawn from a magazine, it will fall into the notch at the point where the staff engages with these discs, and in being moved through the quadrant which it has to traverse in order to pass between the pillar and the opening to the left of the magazine, will carry it with it and so cause each disc to make a quarter circle movement. This movement will be in the direction in which the staff is operated. When the staff is inserted it will be in one direction, when it is withdrawn it will be in the opposite direction. To lock the discs in position three of them are provided with pawls which fit into the notches, and each staff is fitted with lifting pieces which disengage these pawls, and so allow the disc to move whenever a staff is being replaced or withdrawn. There is nothing to prevent a staff being *inserted* at any time, but a check is placed upon its *withdrawal*. To effect this one of the discs is electrically locked. This locking is effected by a detent, or bolt, attached to the armature of an electro-magnet, the cores of which form a closed magnetic circuit, energised by four coils, two of which are in a local circuit and two in the line circuit. The two coils embraced by the local circuit receive a current of a fixed polarity from a local battery. Those in the line circuit are, of course, actuated by either positive or negative currents, as may be sent from the distant station. To lift the lock it is necessary that the current passing through the line coils should be in unison with that passing through the local coils. The cores of the four coils are then fully energised. The local current alone, or the line current alone, is ineffectual; the combined action of the two sets of coils is necessary to remove the lock.

On the right-hand side of the drumhead five switches or contact makers are arranged, two of which are automatically operated by the movement of the discs, and accordingly control the direction of the current passing to or from the line wire. It will be remembered that the insertion or withdrawal of a staff operates these discs, that is, moves them through a quarter of a circle, and it will thus be clear to the reader how readily the insertion of a staff may make provision for the requisite current—positive or negative—for unlocking the instrument at the distant end; and how, on the contrary, the

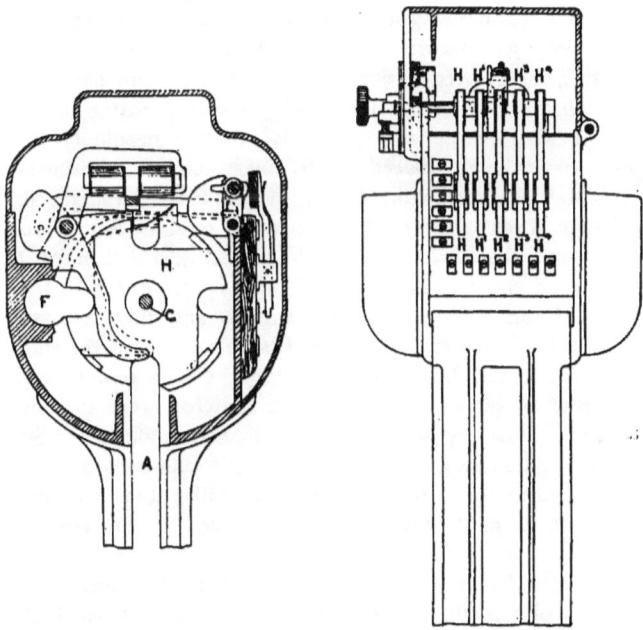

Fig. 74.

withdrawal of a staff will so arrange the switch that the current to be passed to line shall *not* operate the lock.

The remaining three switches are actuated by the right-hand dial and tapper key.

Let us now follow the operation of signalling a train from station A to station B. The signalman at A rings to B by depressing the tapper key. This is acknowledged by B; A then intimates that he requires to take out a staff, and at once turns his right-hand dial to "For Staff." This operation completes the local circuit and energises the local coils at A. B now depresses his plunger key, which sends a current through the galvanometer of his own instrument, along the line wire, through the line coils at A, and through A's galvanometer. A, observing the deflection of the needle of his galvanometer, is aware that the lock on his staff has been removed, and he accordingly lifts and carries the staff through the slot in the drumhead, withdrawing it at the enlarged portion of the slot.

Having withdrawn the staff, A turns the left-hand indicator, thereby registering the fact that a train has been despatched from A to B. The turning of this indicator breaks the line wire circuit, and the galvanometer needle of B's apparatus is thereby placed at zero, by which he understands that a staff has been withdrawn by A and a train despatched.

When B sent a current to A in order to allow A to withdraw a staff, the current sent was in unison with that brought into use on the local circuit at A, and it was owing to the combined influence of the local current and that received from the line wire, that the lock was released and the staff obtained. When A withdrew the staff, the discs were turned a quarter circle, and the automatic switches were thereby reversed. Now any further current which B might thereafter send would be changed in direction, and would consequently be opposed to the current in the local coils; and if A sent a current to B, it would also be opposed to the current in the local coils of B's apparatus, as the reversal of the automatic switches in A's apparatus not only reversed the direction of the current from B, but changed that which A might send, and thus no further staff could be obtained at either station.

If the staff withdrawn from A's instrument is replaced, the discs will be turned back and the connections restored as before; but if the staff is taken on to B and placed in the instrument which is there connected with A, the discs of B's apparatus will be turned back and the automatic switches reversed, which would have the effect of reversing the polarity of the current sent from B, and altering the direction of the current passing through the line coils received from A. Thus both sets of apparatus, that at A and that at B, would be again in unison and admit of the withdrawal of a staff at either end of the section.

Arrangements are made by which, where several trains are required to follow one another in the same direction, and to be maintained a certain distance apart either by an ordinary block system or by an interval of time, tickets can be issued for the trains preceding that carrying the staff, in the same manner as where trains are worked by the staff and ticket system, alone or in conjunction with the block.

A ticket box is secured to the side of the apparatus, and a *ticket staff* is provided which is retained in an independent slot in the column. When it is desired to send say three trains from A to B, the signalman at A obtains permission from B in the usual manner, and instead of taking an ordinary staff he withdraws the ticket staff. With this staff he unlocks the ticket box and withdraws a ticket. The withdrawal of the ticket prevents the ticket staff from being replaced in the column, and no other staff can then be obtained either at A or B. The first train is then despatched with a ticket, and when it has been signalled as having passed the intermediate block post, or if a *time system* is in operation, when the given interval of time has elapsed, the second train is despatched with a (second) ticket. The third train is then despatched with the *ticket staff*, which it carries through to station B.

Provision is also made for switching out intermediate staff stations; for picking up and depositing the staffs at intermediate posts when the train is running at speed; and for locking siding points by means of keys combined with the staff.

It not unfrequently happens that the traffic will gravitate in one direction, and consequently that the tablets or staffs will accumulate at that end of the section. The signalmen are of course aware when this condition attains, and it is for them, or for that one at whose post the tablets are accumulating, to telegraph the lineman in order that he may attend and adjust the difficulty. The number of tablets or staffs is limited, say thirty to each section, fifteen to each instrument. Every tablet or staff, as it is removed from the instrument and passed to the driver, should be entered in the Train Record Book, and its receipt entered by the post receiving it.

When an accumulation takes place at any one post no one but the lineman should be allowed to adjust it. He alone should be able to obtain access to the interior of the instrument, so as to remove from it such tablets or staffs as are required to be returned to the other end of the section.

The transaction should be recorded in the Train Book, together with the time and the number of the tablets or staffs removed. The transaction should also be recorded in a book furnished the lineman, and kept by him, for the purpose. An example sheet taken from such a book will be found in the Appendix.

The lineman should also be provided with a trustworthy receptacle in which to convey the tablets or staffs from one station to another.

CHAPTER VII.

AUTOMATIC BLOCK SIGNALLING.

INVENTIONS for automatic signalling have never been lacking. Unhappily, perhaps, they have usually been the product of the brain of those whose acquaintance with railway working has been limited to the extent of their use of the railway system as an ordinary passenger. Many inventions displaying considerable ingenuity, but of a wholly impracticable character, have been evolved, ending only in disappointment, waste of means and time to their authors. Repeatedly has it been proposed by the aid of treadles, on trains passing over them, to open the whistle of engines as the train is approaching or passing a signal at danger, or to even close the steam valve so as to actually stop the train. To effect this a treadle is arranged alongside the metals, and so connected with the signal that when the latter is placed at danger the treadle is so far elevated that a projection from the engine comes into contact with it. The projection is moved, and this movement is employed to open the whistle or do anything else. Clearly this may be effected under certain conditions, but scarcely those associated with trains travelling at a speed of from 40 to 60 miles an hour; and on many sections of line the speed of the present day exceeds even this. The impact of the leading wheels of a locomotive upon a treadle, or a blow imparted to a projection from such a mass in motion, must necessarily prove distressing to any such apparatus.

Others by the aid of electrical energy propose to establish various modes of communication between signal boxes and drivers, and between train and train. Electrical apparatus, such as can ordinarily be worked by a primary battery, is scarcely suitable for the rough experience to which it would be subject when mounted on the weatherboard of a locomotive engine.

Automatic block signalling has made no progress in England. In one instance only, that of the Liverpool Electric Railway, has it found a footing. Automatic signalling of a certain character, and for special purposes, is occasionally made use of to meet exceptional demands, as for instance in the interlocking of the electric and mechanical block signals; but in no instance other than that previously quoted is it applied to *block* signalling pure and simple. At the same time it must be admitted that a reliable automatic system does possess features, and should be productive of results demanding consideration. Under its employment the line may be divided into just such sections as would most readily meet the demands of the traffic: there need not be that arbitrary division which has now to be observed in order to meet the demands of local sidings, stations, &c. Although, naturally, some cost must attend its upkeep, the saving in signal boxes and wages would be considerable. In America the objections to its employment that attain in England do not exist; but we must not lose sight of the fact that in England, as a rule, the block sections are not of the same extent as in America, and that in the majority of cases a signal box, presided over by a signalman, is necessary quite independent of the block signalling; that is, it would be required to be there for working points, cross-over roads, &c. Given the need for the signal box, very much of the advantage of an automatic block disappears. The block sections may not be so regular or so well adapted to the traffic, but there is the presence of an officer to deal with any irregularity in the traffic or the working gear which may arise. We cannot question that personal responsibility carries with it greater confidence, greater reliability, than can any automatic arrangement.

Let us consider for a moment the mass of traffic passing over some of our trunk lines worked under the most reliable automatic system, one worked upon that principle which shall entail, in case of failure of any one portion of the system, that the signals shall assume the danger position. Train after train would of course have to be stopped or slowed until the derangement were adjusted. What would this mean to a main line traffic? Probably disorganisation of the entire train service for the day, or even longer. When we reflect that under our present block system the traffic is conducted at the rate of over 80,000 train miles to one failure in connection with the block signalling apparatus, we are tempted to ask if such a result could be anticipated from automatic signalling.

Another forcible factor is the power required for operating automatic signals. Even with encased signals, i.e. signals operated within a glass-faced case, the power required to satisfactorily work them would severely tax any primary battery. Again, we must not lose sight of the possible effects of our climate upon such a protection. Moisture would gather upon the glass, snow and ice accumulate, all tending to obscure the condition of the signal from the view of the driver.

The foregoing observations are, it will be observed, on the whole unfavourable to automatic block signalling; still the suggestion will assert itself—whether something of the kind might not find employment upon branch lines where stoppage of traffic might not be attended with such serious consequences as on a busy line; or if it might not be so associated with the existing system of block signalling as to admit of its employment at intermediate points between signal boxes some distance apart. There are, of course, many signal boxes established solely for signalling purposes. If in place of these a system of automatic signals, the action of which were, by electrical repetition, brought under the observation of the signalman on either side, there would appear to be reason in believing that such might be employed without risk and with some economy.

LIVERPOOL OVERHEAD ELECTRIC RAILWAY.

The automatic block system employed on the *Liverpool Overhead Railway* is the invention of Mr. I. A. Timmis. The line is divided into block sections in the usual manner, in extent about half a mile each. At the commencement of each section there is a station. This, the sections being so short, admits of the home signal for one section and the distant for the next being at the same station; the former serving to admit the train into the station and protect it while standing there, and the latter as starting signal to allow it to proceed into the next section. The last vehicle of each train is fitted with a device for making contact with a lever projecting from a box arranged for the purpose on the permanent way, the object of which is to make or break the electric circuit by which the signals are operated. The contact surfaces are long and move over each other with considerable friction, which ensures good contact.

The signals which govern the trains are much smaller than those in general use, on account, it is said, of the short distances between stations and the low fixed speed of the trains. What is termed by the inventor a "long-pull electro-magnet," is employed to operate the signals. The energy required is 5 ampères at a pressure of 40 volts (200 watts). On the signal being placed in the line clear position, the current operating the magnet is reduced by the automatic insertion of a resistance which brings it down to ·25 ampère. The electrical contacts on the starting signals are formed by means of a vessel containing mercury, which, when in one position, connects two contact plates, and when in the opposite position, leaves them disconnected. Provision is made for dealing with trains which, owing to some irregularity, may require to cross over from one line to the other and return to the station from which it started. This is effected by taking the main circuit through a switch connected with a point lever, which, when worked, breaks the circuits of the two lines, while the parts are at the same time so connected as to admit of the due operation of the signals for the roads open to traffic.

Fig. 75 shows diagrammatically the arrangement of a station, the signals and circuits. We will assume a train in progress between stations 2 and 3, and approaching the latter. The home signal at A is down. Just before the train clears the signal it passes the breaking contact B, which sets the signal to danger. At B there are two sets of contacts broken in succession to remove all chance of failure. With the home signal at danger, the train reaches the platform of No. 3 station. On the starting signal C being lowered the train moves on, and, as it passes the signal, puts it again to danger by means of the contact box at D. Some two hundred feet further on is "making-contact" E, which; when operated by the passing train, lowers the advance starter at station No. 2 and the home signal at No. 3. The circuit of the latter is from the contact E, through mercury contact on the starter arm C, along the wire to the home signal A, through the magnet and the breaking contact B, to the battery on the

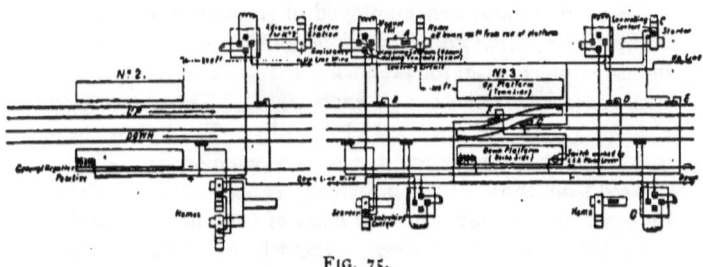

FIG. 75.

down platform, and thence back to the contact E. Immediately the signal is pulled off, a new circuit is set up by the contacts in the mercury box on the signal post. The new circuit passes from the breaking contact B through the magnet coil and the resistance to the battery, and back to the contact B. The same course of action is pursued as the train passes on to station No. 4.

The circuits for the cross-over road are also shown on Fig. 75. On the down platform there is a switch which is locked by a key, which key is also necessary to release the point lever, and which must therefore be turned before the points can be set. The key is kept in a box on the home signal post, and when it is removed the signal goes to danger automatically, should it not already have been placed in that

L 2

position. By this means the electro-magnet which works the signal is not interfered with, and when the cross-over road is restored for through working, and the key is taken out and replaced in the home signal, the electrical condition of the signal is reinstated.

It is impossible for any train to travel without at least one signal, the "starter" being at "danger" behind it, owing to the lowering circuit passing through the contacts on the "starter," and to the fact that the circuit is only complete when the "starter" arm is at "danger."

ELECTRO-PNEUMATIC BLOCK SIGNALLING.

An automatic system employed in America to some extent is the *electro-pneumatic*. Its application involves, as the title indicates, mechanism for producing the necessary air pressure, with *receivers* for the use of each signal. The pressure and exhaust valves are operated electrically. The line is cut up into the necessary sections, the rails of each section being insulated from those of the adjoining section. This is effected by separating or insulating the ends of the rails and employing fish-plates made of wood. About fifty feet to the rear of each section is a post carrying two semaphore signal blades. The upper one is the stop or *home* signal; the lower one the *distant* for the *home* signal applicable to the next section. A suitable primary battery is arranged with one of its poles to one rail and the other pole to the other rail. On each signal post a relay is arranged for the purpose of bringing into play a powerful battery for operating the pneumatic valves. Fig. 76 will explain the electrical arrangement. A is a portion of a section, B and C are entire sections. It will be observed that at *b* a battery is in circuit with the rails, one pole being connected to one rail, and the opposite pole to the other; and that in the neighbourhood of the signal post commanding section B, a *relay* is also in circuit with the rails. It will be clear that so long as the rails afford good continuity, a current will flow from the battery *b* along the rails and through the relay, the armature of which will be drawn over to the contact stud in connection with the pneumatic signal cylinder. The exhaust valve of *d* will now be closed and the pressure valve opened, and the arm governing that section will be held *off*.

Now on a train entering the section the battery *b* will be short-circuited. The current will be cut off from the relay, and the pressure valve of *d* will be closed and the exhaust valve opened. The arm being no longer held off will, by gravity, rise to the danger position, and remain so until the train has passed out of the section. Any means by which the battery *b* might become short-circuited, or any means by

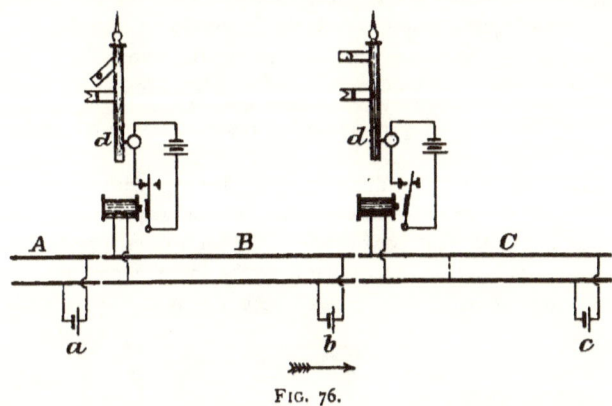

FIG. 76.

which the current from *b* might be prevented from traversing the coils of the relay, would result in the signal falling to danger.

Each signal post is, as has already been said, fitted with two arms, or blades, the upper being the *home* signal for the section which it commands, the lower being the *distant* signal for the section in advance. On the train entering section B, the *home* signal arm is raised to the danger position, as previously explained. The lower arm—the *distant* for section C—is electrically connected with the upper arm, which in moving to the danger position causes the lower arm also to assume that indication.

On the train passing into section C, the battery b again comes into operation, and the *home* signal for B is again lowered to the *all right* position; but not so the lower arm, that remains at danger as an index of the home signal for B. At the same moment battery c is short-circuited and the home signal, as well as the distant for section C, are raised to danger. On the train entering section D, the distant for B will be lowered, but not that at C. Thus it will be seen that each train has two distant signals and one home signal to protect it. The operation of the distant signal is consequent upon that of the home signal, the operating power being compressed air, the controlling power electricity.

The diagram assumes a train in section C, at the point indicated by the dotted line connecting the two rails.

THE HALL SYSTEM.

One of the chief difficulties in dealing with automatic signalling is that of battery power. On railways or tramways worked by electrical power there will be ample energy, because that energy can then be drawn from the source which provides the means of locomotion; but where this is not available the

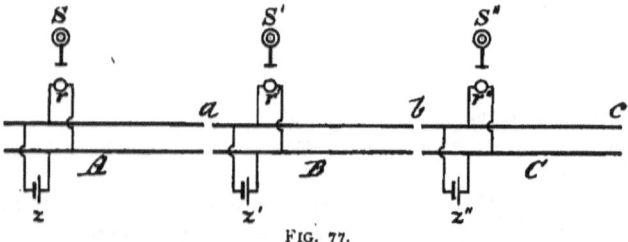

FIG. 77.

apparatus must be constructed to work with the power to be derived from a primary battery. Still there are systems in operation, actuated by these feeble currents, which have worked apparently with satisfaction. Such a system is that known as the *Hall system*, largely employed on the United States railways.

This system is probably fairly well known. Its establishment entails the division of the metals into sections, that is, their insulation one from the other at given points, each forming a block section. Thus, assume Fig. 77 to represent a line of rails divided into three sections A, B, C, the rails being separated and insulated at a, b, c. Now it is clear that if we attach a battery to the rails of each section, as shown at z, z', z'', and a relay as shown at r, r', r'', capable of operating an enclosed signal, S, S', S'', on a train entering upon and traversing any one of the sections, it will, so long as the wheel upon one of the rails is in metallic connection with the other rail, short-circuit the battery; and if the relay is so arranged that it shall, when the current is flowing, hold the signal off, and when no current is flowing, exercise no effect upon it, and that the signal shall then under the influence of gravity fall to danger, we have a means of controlling the signal perfectly in accord with the movement of the train.

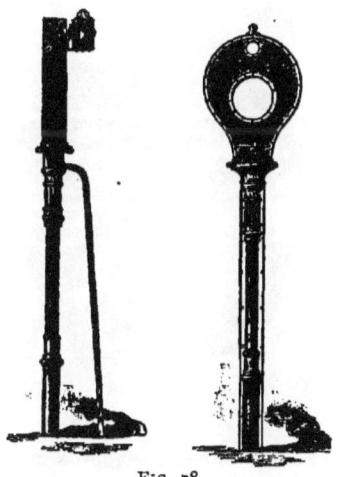

FIG. 78.

This is the principle upon which the "Hall system" is based, and it will be clear that, provided we can ensure the necessary isolation of the two metals which constitute the two poles of the battery, and satisfactorily insulate the one section of rail from the next, we may operate as safely as is reasonably possible, a set of signals for the control of trains.

The form of signal in use is shown in Fig. 78. It consists of a metal or wooden hood, mounted upon a pillar so as to be clearly seen by an approaching train. In the centre of this hood is a large circular orifice covered with glass both back and front, and above it is a smaller orifice similarly protected. Inside the case, Fig. 79, is an electro-magnet, with two discs formed of a light textile material coloured red, mounted one upon each end of the armature. One of these discs is large enough to fill the larger orifice, the other is sufficient to fill the smaller one. There is a proper balance between the two, so that there is just sufficient tendency to fall to that position which will cause the discs to fill the orifices when the armature is unaffected by the coils. The position of these discs when under the influence of the current is shown by the figure, which presents the open orifices clear, which to the driver is "line clear."

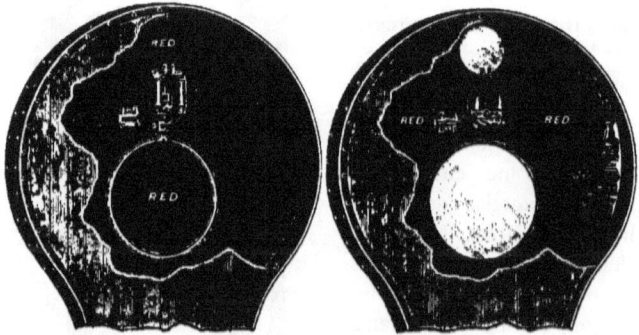

Fig. 79.

The relay employed is one of an ordinary character, capable of being actuated in one direction only, the electrical arrangement of the signal being that indicated by Fig. 80.

So far nothing of the kind has been successfully instituted in England, probably owing, in the first instance, to that objection which exists to the employment of anything which fails to carry with it a personal responsibility in the movement of any railway train; and, in the next instance, to the difficulties which would possibly attend its application. The reader will readily associate in his mind the numerous and special provisions which would be necessary in dealing with such a system, with our numerous points and crossings; but, of course points and crossings must exist on the American lines, not so numerous probably, as their wayside stations are not so frequent as in England ; still, this system is reported as having worked all the traffic of the Illinois Central Railroad, in and out of Chicago during the Chicago Exhibition. There must have been then no lack of such conveniences. Points and crossings would of course require special treatment ; they would form small sections, each operating signals for all roads fouling such crossings, and at, the same time actuating

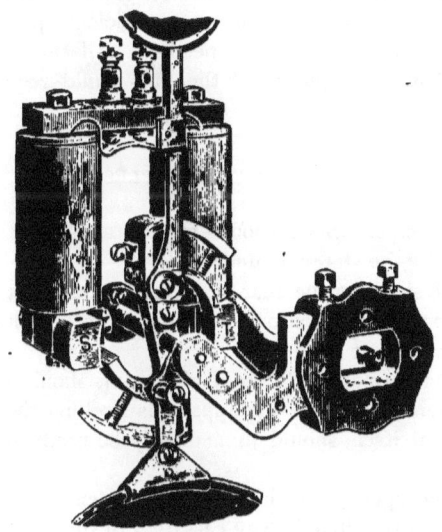

FIG. 80.

"ground" signals as well as the signals protecting the section of which the crossing formed a portion. Any number of signals may, of course, be operated from the same relay, and those signals may be placed in any position required by electric wires carried to those points.

It will, in conclusion, be noticed that there still exists as a difficulty to be overcome, results arising from the humidity of our atmosphere, which, as a rule, will maintain the ballast between the rails in such a condition as to induce a certain amount of cross connection, thereby exhausting the battery, and possibly setting the signal at danger. With the rails electrically connected, or "bonded" one to the other, so as to diminish the electrical resistance of the section, and the *chairs* set upon creosoted sleepers, it is still probable working may, even with a British climate, be satisfactorily assured. There then remains the insulation of the continuous rail. In America, hard wooden fish-plates have been used. Possibly, however, the better plan would be to terminate each rail upon an independent chair, allowing just sufficient space between each to insure the insulation needed. All this is a matter about which the engineer of the line would require to be consulted.

GOVERNING PRINCIPLES.

The principles which should govern automatic signalling may, broadly, be stated as under :—

(i) The signal in the rear of a train should not be taken off until that for the section which the train is about to enter has been placed at danger.

(ii) The normal condition of all signals should be danger, and any derangement of the apparatus working the signal, or of the signal itself, should immediately be productive of this signal.

(iii) The agency by which the "line clear" signal is produced should be active. Any cessation of this agency should result in the signal going to danger.

(iv) The design and arrangement should be such that the "line clear" indication cannot be produced by earth currents or lightning.

(v) The apparatus should be so designed that on emergency it may be operated by hand.

(vi) Preferably it should conform to the system of signals in use, both with regard to form and mode of operation.

(vii) And it should be strong, simple, not easily deranged, and readily repairable.

CHAPTER VIII.

INTERLOCKING.

WE have now, under the influence of the Railway Improvement Act, reached that position when our entire railway traffic may be said to be worked under the protection of the block system. Let us consider what this means.

The line of railway is divided into sections. Each section is provided with signals for the guidance of the engine driver. These signals, which we may denominate mechanical signals, are worked from a signal box, in which are arranged certain electrical instruments which are provided for the purpose of telling the signalman whether the section of line between his post and that on either side of him is clear or not.

With very few exceptions the two systems of signals—the electrical and the mechanical—are worked quite independent the one of the other. It is quite possible for a signalman, mechanically and in absence of mind, to pull off his starting, or home signal, and so allow a train to pass into a section which the electrical signal tells him is blocked. When we consider how many thousands of miles of line are worked in this manner, it must be admitted that the comparative freedom from accident which characterises the working of British railways, reflects upon those responsible for the manipulation of the signals the greatest credit.

Although the condition indicated above exists to so great an extent, it must not be thought that efforts to remedy it have been wanting. The desirability of relieving the signalman of that great responsibility which it casts upon his shoulders, and of rendering it practically impossible to operate the mechanical signals adversely to the electric signals, has been recognised, and means of doing so provided; but only to a slight extent have these means been made use of, and that, mainly, very recently.

The earliest effort to interlock the mechanical with the electrical signals was, it is believed, that brought out by Mr. W. H. Preece in conjunction with the author, in 1870. The principle is fully described in the Patent Specification No. 1268, 1870. The first instrument fixed was at Southampton, followed by others at Wimbledon, on the London and South Western Railway. The interlocking, in this instance, was confined to rendering the controlling mechanical signal dependent upon the action of the electric signal. There was no control of the electric signal. A more complete system, in which this was effected, was subsequently laid down on the Kew branch of the same railway. Circumstances at that time called both Mr. Preece and the author away from the railway service, and the matter, so far as they were concerned, was allowed to drop.

Their efforts were, however, soon followed by Messrs. Saxby & Farmer, under Hodgson's Patent, and at a later date by Sykes, Spagnoletti, and again by the author, in his system applicable to the existing form of block instruments. The later form of interlocking produced by Mr. Sykes has been largely applied by the London and South Western, and the Great Eastern on their metropolitan and main lines entering London.

The requirements of an interlocking system, by which the mechanical system of signals is combined with the electrical system, are :—

(*a*) That it shall not be possible to lower the *key* signal —i.e. say the starting or home signal—unless the electrical block signal indicates *line clear*.

(*b*) That at any time it shall be in the power of the signalman to place his mechanical signals at danger.

(*c*) That on the electric block signal being placed in the *train on line* position, it shall become locked and remain so until released by the passing of the train out of the section.

(*d*) That it shall be in the power of the signalman at any moment to place his block instrument at *line blocked*.

(*e*) That the removal of the lock (clause *c*) on his instrument by the passing of a train shall not of itself relieve the "line blocked" signal, but merely prepare the parts of the instrument, so that the signalman may himself release it.

(*f*) That the design or the arrangement of the apparatus shall be such as shall enable the passing train automatically to place the "key" signal at danger, or render it obligatory on the part of the signalman to do so, and thus ensure its again becoming locked before he can give the *line clear* signal for a following train.

The object of these provisions is so clear that comment is scarcely needed. It is, of course, desirable the signalman should at any moment be able to throw all his signals on, and in like manner to block his roads by placing his block instruments at danger. The interlocking should in no way prevent this, or be of such a character as to impede his action.

There can be no question of the desirability of interlocking the mechanical with the electrical signals; the only marvel is that it has not been done to a larger extent than is the case. Possibly, although no doubt cost is a potent factor, railway companies have hesitated to replace a well-understood block system by one which would call for certain changes in mode of manipulation as well as in the character of the signals themselves. This, however, does not, as will be seen, apply to all the systems which have to be referred to. Cost is, possibly, the main reason; and it must be conceded that the advances made by railway companies for the security and more complete regulation of traffic have, within the last twenty years, been of a highly marked character, and must have entailed, and still entail, serious outlay.

It is here that the question arises—Is that which is now being done—the mode of interlocking at present pursued—complete? Does it meet all that is required? Considerable expense is being incurred. Is there an approach to finality about it? It must be admitted that even as we see things at present, it is not finality; it does not meet every contingency.

It provides for a great deal, but it does not provide against a divided train. It may be argued that the signalman ought to see that the whole of the train has passed before clearing it. But he *may not*. The loophole is there, and it is desirable that in adopting interlocking every possible contingency should be provided for. The day, of course, will come when the deficiency will present itself in no unmistakable manner, and then much previously done will be thrown away, and finally that which ought to have been done in the first instance will have to be adopted.

The course now pursued provides for releasing the locked block instrument by the passing train operating a treadle or other device in connection with the railway metals. The depression of the rail by the weight of the passing train or engine, causes a current to be sent through the locking part of the block instrument by which the lock is released. This is effected by the first vehicle—the locomotive.

Now this is manifestly wrong. The release of the lock should be dependent upon the passage of the *last* vehicle.

There should be no difficulty about this, nor should it be more costly than the present system. All that is necessary is that the last vehicle shall carry a contact maker, by means of which the electrical circuit may be completed. And in order that there should be no omission in applying it to the last vehicle, and again releasing it when that which, for a portion of the journey, has formed the last vehicle is no longer such, the arrangement adopted should be contingent upon the tail lamps. This means that on the tail lamps being removed from a carriage or van, the contact bar should be released, preferably automatically; and *vice versâ*, when the lamps are placed in position, that the contact bar should assume its proper position for signalling the passage of the train.

THE SINGLE-NEEDLE BLOCK INTERLOCKING SYSTEM.

In designing this system the inventor has recognised the importance of making it not only applicable to the existing form of block instrument but that it should, in other respects,

to Railway Working. 159

conform to the mode of working. The final action, which produces the *train on line* signal, locks the instrument in that position until the lock is released by the train as it passes out of the section. The action of the signalman and the course of the signalling is in no way interfered with. The lock on the *key* signal acts by gravity, or the influence of a spring, and is released only when the block instrument renders the *line clear* signal.

The apparatus consists of three parts: the block signalling instrument; the lock on the *key* signal; and the rail contact maker.

Fig. 81 shows the block instrument in elevation. Ordinarily the commutator handle A is, as has been previously explained, pegged over to the positions rendering the signals *train on line* and *line clear*. We have now, however, to employ the small pedal levers, *d d'* provided on either side of the handle for that purpose. Reference to the vertical transverse section, Fig. 82, and to the enlarged view, Fig. 83, of that portion of the apparatus by which the commutator handle becomes locked over to the *train on line* position, will make clear the reason for this, as well as render the general action of the different parts intelligible.

FIG. 81.

On the handle A being carried over into the position B in order to render the *line clear* signal, the pedal *d'* is pressed down. This action of *d'* causes a small catch piece to enter a notch cut in the flange of the socket of the instrument handle, and so retains the handle in that position till released by the signalman again raising d'. In like manner, in order to render the *train on line* signal the handle is placed in position C, and the lever d is pressed down. It also connects up and brings into play certain additional battery power for operating the lock on the starting or key signal at the distant station. d has something more to do than d'. It is connected by a crank g with the circular inclined plane f, which is normally retained in the position shown in Fig. 83 by the spiral

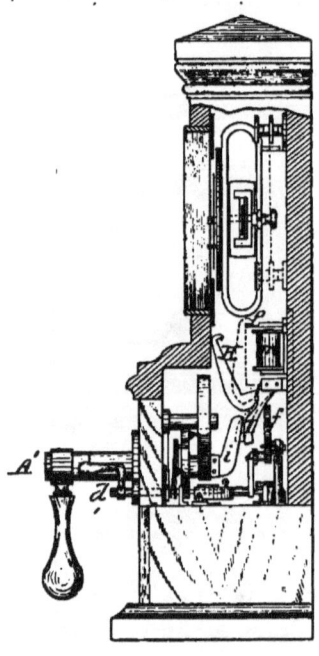

Fig. 82.

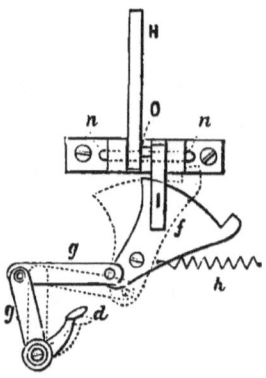

Fig. 83.

spring h. d on being pressed down causes f to move forward into the dotted position, and in doing so to lift the locking part I, which, in its turn, operates H so far as to cause it to engage with the catch piece P of the armature of the small coils m. To the end of the shaft A' is fixed, rigidly, a steel bar i. The locking piece I has, by the action of d, been carried past i, and at the same time elevated to such an extent as to cause it to project slightly in front of i, as will be seen in Fig. 82. If, now, the handle A is released by the operator, it will be found to be held fast—the two locking pieces I, held in position by the catch piece H, having engaged with i so as to prevent the latter passing the former. This condition is maintained until H is released by the armature P of the small coil m being depressed, as would be the case on the passage of a current through these coils.

The small electro-magnet m is in circuit with the rail contact maker, and on a train passing over it a current is passed through m, the armature P is attracted, the catch piece H is released, and the lock can be removed by the signalman urging the handle A so as to allow I to fall away to its position of rest, clear of i. The locking piece I is loosely connected by a pin o with the catch piece H, so that when I is lifted it has a certain initial movement before it affects H, and so that H may have a like independent movement when falling towards I. This is accomplished by allowing the pin o to work in a slot cut in H. The two pieces H and I have thus a limited initial movement independent of each other.

The signal lever locking instrument is shown in Figs. 84 and 85, the latter in plan and the former in vertical section. The crank end of the signal lever R is connected to a slide bar S. The slide bar moves in a groove provided in the iron frame T, and at V a notch is cut in it, which notch coincides with a transverse slot cut in the sides of the groove which S traverses. W is an electro-

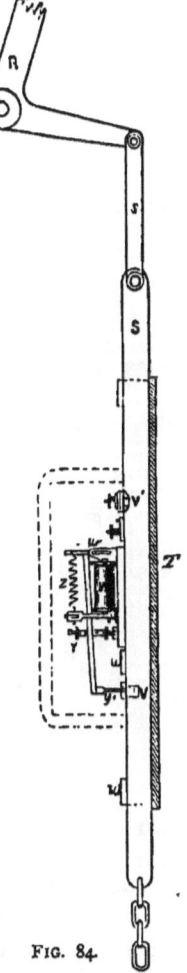

FIG. 84.

M

magnet, to the armature of which is attached an extended lever, having at its extreme end a small cross-bar y' of steel. The rectangular cross-bar y' exceeds in length the thickness of the slide bar S, so that when residing in the notch its ends may, if urged, rest against the slot cut in the sides of the groove, and prevent the slide bar from being moved. When this is the case the slide bar and, consequently, the signal lever are locked.

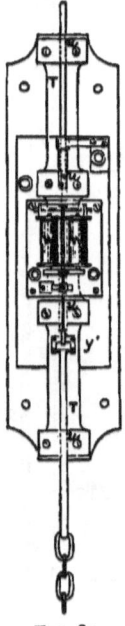

FIG. 85.

In order to oblige the signalman to place his lever in the position necessary for locking it, a notch V' is cut in the slide bar, and an insulated spring is attached to the frame in such a manner that unless the lever is in the required position the spring will not enter the notch to complete the battery circuit which is necessary to enable the signalman to give the *line clear* signal for a following train to come forward.

In dealing with Figs. 81 to 83, it was pointed out that on pressing down pedal d', it not only locked the handle A over but it also increased the battery power. The coils of the block instrument are capable of being worked satisfactorily by two Leclanché cells. The electro-magnet W operating the lock on the slide bar, Figs. 84, 85, is wound with a much thicker wire than are the coils of the block instrument, and it requires a larger force than is necessary to work the block needle, that it may attract the armature and so lift the locking piece. Thus, for the ordinary working of the block instrument only two or four cells are required; but when rendering the *line clear* signal the current must be increased to such an extent as will lift the lock on the slide bar, and this, as has been explained, is effected by, on pressing down the pedal which locks the needle to the line clear side, bringing into circuit the necessary additional battery power.

In order to make quite clear the operation of the several parts, we will follow out the signalling of a train. On a train being about to start, the signalman at the distant end of the section is asked if line clear. If such is the case he fixes the handle of his block instrument over to produce that indication. In doing so he has pressed down d', locked his handle A to line clear, sent the increased battery current to the starting station, lifted the lock from the starting signal, the signal is lowered, and the train starts. Its departure is now notified, and the distant station then releases d', turns his handle A over to the opposite position, presses down d, and by its means locks his handle to *train on line*. As soon as he released the *line clear* signal, the current flowing through the electro-magnet W became greatly reduced, leaving the lock on the starting lever ready to fall into its slot as soon as the lever working the signal became restored to the danger position for the protection of the train just started.

On the train reaching the far end of the section it, in passing over the rail contact maker, operates the locking coil in the block instrument, and the lock which has *retained* the *train on line* signal during the time the section has been traversed by the train is now released, and the signalman is at liberty to restore the needle to its vertical position, indicating *line blocked*, or, if necessary, and he has placed his starting signal at danger for the protection of the train just arrived, to give the *line clear* signal for a following train.

Messrs. Siemens' "hydrostatic" contact maker is employed for signalling the passing of the train out of section, and thereby releasing the lock on the block instrument.

This contact maker depends for its action upon the bending of the rail due to the passing of a train. It is fixed to the rail itself at points 2 feet 8 inches apart. As soon as this portion of the rail is weighted by the wheel of a passing train, a very slight movement is obtained by the bending of the rail, which is transferred to the electrical contact by hydrostatic pressure.

Fig 86 is a perspective view of the complete apparatus, Fig. 87 a side view of the same attached to a rail, and Fig. 88

is an enlarged section, showing the working parts of the apparatus.

The contact maker, seen in the centre of the block, is supported by a strong cast-iron beam, Fig. 87, M, L, L¹, M¹, which, by means of claws and screws, is firmly screwed to the bottom flange of the rail. In the centre of the beam is a broad

Fig. 86.

shallow dish, the rim of which is covered by a thin steel plate, *b b*, Fig. 88 (similar to the diaphragm or membrane of a telephone, but larger). On this membrane rests an iron disc *c c* which is held at the centre by the plunger *d*.

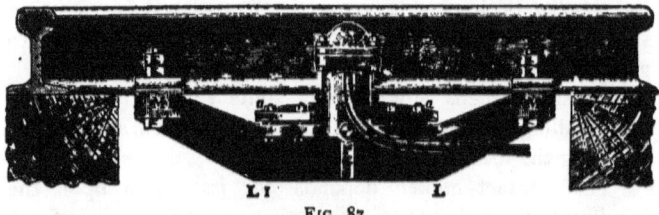

Fig. 87.

This plunger is so adjusted that it just touches the under surface of the rail when the apparatus has been screwed firmly to the latter.

The dish, steel membrane, and the disc, are covered by a dome *a a*, in the centre of which moves the plunger *d*, and on this cover and under the foot of the rail is a solid ring of india-rubber *t*, laid round the plunger *d* to protect it from the penetrating sand which, otherwise, might impede its free action.

A second and higher vessel, G, Fig. 88A, communicates by a narrow tube with the space under the membrane *b b*, widening upwards into a cup *r*. That part of the tube which projects into the cup *r* is made of insulating material; there is besides a small hole *h* in this tube and another *s* at the bottom of the cup, thus establishing free communication between the interior of the tube, the cup *r*, and the upper vessel.

The hollow of the shallow dish under the diaphragm *b b*,

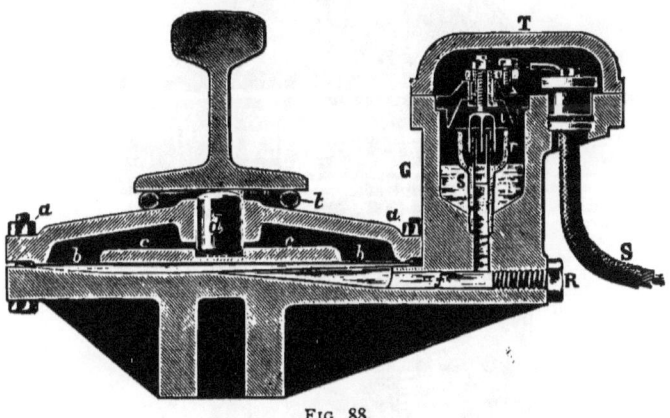

FIG 88.

Fig. 88, and the upper vessel are filled with mercury till the bottom of the cup *r* is just covered. In consequence of the difference of level in the upper vessel and the dish, the diaphragm *b b*, and with it the plug *d*, is always pressed firmly upwards against the foot of the rail by a hydrostatic pressure of about 66 lbs. As soon as the rail experiences a slight deflection between the points of attachment M, M^1, due to a passing weight, such as a railway train or wagon, the foot of the depressed rail presses against the plunger *d*. This pressure is communicated by the disc *c c* to the diaphragm *b b*, and thus to the mercury confined in the dish, and as the ratio between the surface of the mercury in the dish and in the tube is as 600 to 1, a considerably increased motion is imparted to the mercury in the channel, forcing it up the tube and rapidly filling the cup *r*. As soon as the train has passed, the deflection of the rail ceases and the plunger *d* is released. The mercury then runs slowly out of the cup through the side hole *s* into the vessel G, and thence through the small hole *h* into the dish under the steel diaphragm *b b*, so that any mercury overflowing the top of the tube will fall into the chamber of

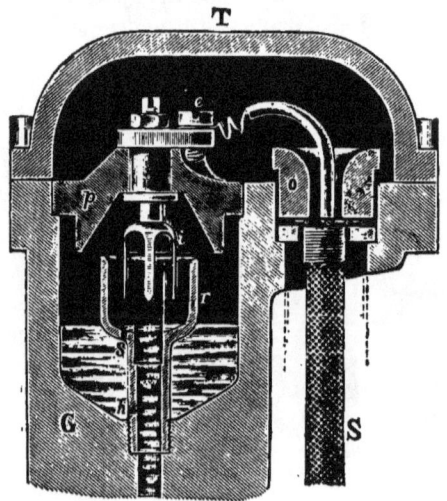

FIG. 88A.

the vessel G, and thence pass back into the channel and main vessel when the pressure on the diaphragm is removed.

The adjustable contact fork *i* projects into the cup *r* and the insulated end of the tube without touching either of them. This fork is connected to the cable S by a terminal screw *e*, attached to a slate top piece *p* (see Fig. 88A), thus being insulated from earth except when the column of mercury rises and the fork and mercury are brought into contact. As the iron dish, the dome cover and the rail are all in contact with earth, the dipping of the fork i into the mercury closes the circuit and actuates the signal.

All the parts coming in contact with mercury are made of iron. Above the slate top piece p, to which the fork is fixed, is a cast-iron lid T, which covers the junction of the cable S and the entire contact-making parts.

By screwing the contact fork i higher or lower, a shorter or longer duration of contact can be obtained.

The mercury employed for the mercury vessel must be pure.

SAXBY AND FARMER'S INTERLOCKING SYSTEM.

Messrs. Saxby and Farmer, the well-known railway signal engineers (working under Hodgson's patent), were not slow in recognising the importance of combining the electric with the mechanical block signals.

In appearance the instrument employed is practically that of Tyer, with, of course, a different form of commutator, it being necessary that part of the apparatus should be so designed as to lock itself when operated in the train on line (or line blocked) direction. These instruments may be worked on the constant-current system, in which case three wires are needed, or by the momentary current, which requires but one line wire.

The outdoor starting signal is provided with an *electric slot* which obliges the starting signal to be at *danger* so long as the electric block signal is at *line blocked*. Provision is made that the train itself shall place the starting signal at danger.

Messrs. Saxby and Farmer claim the following as advantages attending the use of their system.

(*a*) One wire works both audible and visual signals of the most reliable kind, for both up and down lines; the electric current is transmitted by simple means; there being only one handle carrying the spring plunger, there is no liability of mistake.

(*b*) Perfect control of outdoor signals exhibited to drivers,

168   The Application of Electricity

which signals can only be worked in conformity with the Electric Block indications.

(c) Signal invariably put up to " Danger " behind trains, as two men can put it up, and if they fail to do so the train itself does it, and then the signal cannot be lowered again until again released by the man at the next station in advance.

(d) Once " Line clear " has been sent and a train has entered the section, " Line clear " cannot be sent again until the

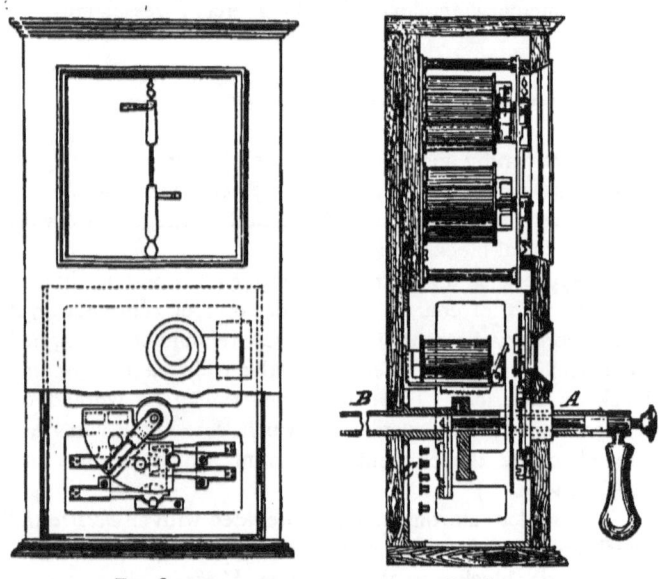

FIG. 89.   FIG. 89A.

train leaves the section, and in doing so it unlocks the handle of the instrument there.

(e) Interlocking mechanically the electric block instruments with the point and signal levers so that messages of " Line clear " cannot be sent and levers used for shunting at the same time. After " Line clear " has been sent, and a train has entered the section, the points (such as should not ·be moved) remain locked until the train has passed.

(f) Economy of battery power, the connection to battery

being cut off directly work is done, and no battery in use to slot till spring catch of lever is grasped for the purpose of working the signal.

A characteristic of these appliances is that each can be employed independently of the other, or all may be used in combination.

Two block instruments are required in each cabin, one for traffic to and from the next station in one direction, the other for traffic to and from the next post in the other direction. These instruments, Fig. 89, are fixed on a shelf immediately over the lever frame, and are connected there with a rod extending from the block instruments to the lever frame. These rods are connected by a crank with the shaft of the commutator of the block instruments, and similarly with a locking piece or bar in the frame (Fig. 90).

On the line of railway are arranged electric contact makers, or *treadles*, for restoring the signals to *danger*, and for releasing the commutator handles of the block instruments.

The handles of the instruments are designed to occupy three positions, Fig. 91 — (i.) normal towards the left, indicating *line blocked*; (ii.) towards the right, indicating *line clear*; and (iii.) vertical, indicating *train in section*.

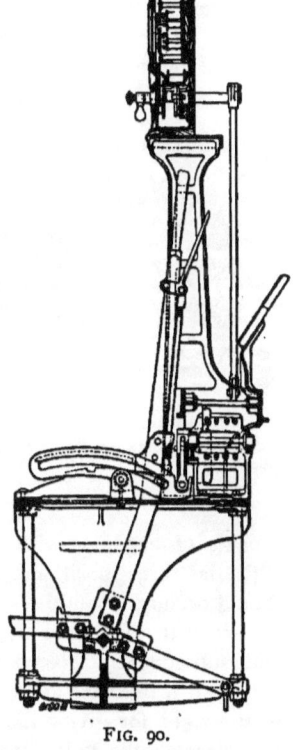

FIG. 90.

The handle of the instrument, Fig. 89A, is fixed upon a spindle or shaft A, which at its far end B is attached to a crank in connection with the rod leading to the signal frame.

This shaft is tubular, and carries within it the rod of a spring plunger, the knob of which is seen to the front of the handle. This spring plunger, when pressed, impinges upon certain contact plates C arranged within its field, and thereby causes positive or negative currents, according to the number of times it is pressed, to be sent to the adjoining station. The character of the current is, of course, determined by the position of the handle, which is maintained in the position required by means of the sector plate N and the spring seen immediately below it, and which, as will be observed, is

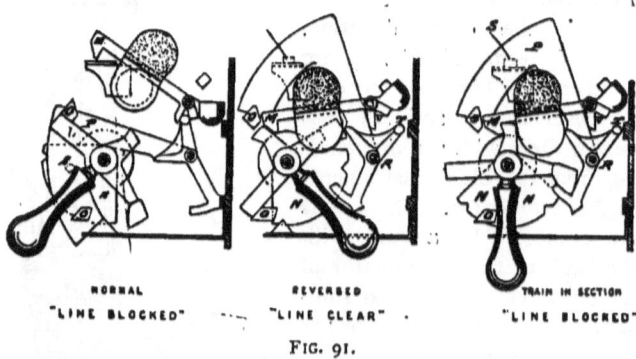

NORMAL REVERSED TRAIN IN SECTION
"LINE BLOCKED" "LINE CLEAR" "LINE BLOCKED"
FIG. 91.

capable of engaging with the former at three points, uniform with the three positions which it is designed the handle should occupy.

When the handle is moved over to the *line clear* position, and subsequently placed in the middle or vertical position, (Fig. 91), it becomes locked, so that it cannot be moved either to the right for giving *line clear* a second time, or to the left for releasing the point levers which may be mechanically interlocked with the handle. This interlocking of the handle is effected in the following manner:—

Inside the instrument, upon the shaft or spindle which passes through it, is fixed the three-armed lever L, the straight arm of which is provided to limit the movement between the stops O; and the hooked arms are for the purpose of engaging with the hooks or lock pieces P and M, provided for the purpose of securing the handle in position. The curved plate P is pivoted on the frame of the instrument at R, and is raised by the movement of the handle from the "normal" to the "reversed" position; the hooked limb of L, acting near the fulcrum of P, lifts it into the positions shown in Fig. 91. Upon the inner side of P there is a small stud S which is bevelled at the top and is square at the base. When P is raised by the action of L, the stud S presses against the armature of an electro-magnet, maintained in its normal condition by means of a spring, and pushing it aside, passes over and rests upon the top of it. P now rests upon the top of the armature, the hooked end of P engages with the hooked limb of L, and the raising of P allows M by gravity to fall and engage with the other arm on L. M carries two plates, the upper one of which is coloured red, the lower green. These are seen through an aperture provided in the screen of the instrument, according to the position they occupy, and so indicate to the signalman the position of the plate P. When P is down, as in Fig. 91, the handle is free, and green appears in the aperture; when the handle is vertical and locked, red is shown.

The handle, when locked for *train in section*, can only be released by a train passing over the releasing treadle, which, completing an electrical circuit between the treadle and the block instrument, causes a current to pass through the electro-magnet previously referred to, when the armature of the same being attracted, the stud S is no longer supported by it, and by its own weight P falls to its normal position. This movement of P causes a pin T at its extremity to enter the V aperture of M, and pressing upon its forward limb again raises it into that position which brings the green disc before the aperture of the block instrument, indicating that the lock has been released.

To one arm of the plate P is fixed a platinum point, which, when the plate is raised, descends into a cup of mercury; the object being to break the treadle circuit when the plate is down, and to make it when the plate is raised.

The Application of Electricity

The *releasing* contact maker in connection with the rail is shown in Fig. 92. A is the railway metal ; B a lever operating the electrical contact ; D a rod connecting this lever to a box partially filled with mercury, and which box is free to work on a centre F. On a train passing over A, the flange of the fish plate presses upon the free end of B ; the mercury box E is tilted up, and the mercury is moved to form contact with two platinum wires which are connected with the cables, thus completing the electrical circuit and operating the armature of the electro-magnet previously referred to.

The *electric slot* on the starting signal is illustrated in its several positions in Figs. 93A, B, C. The vertical rod R moves with two parallel links r' r", which are pivoted on the main

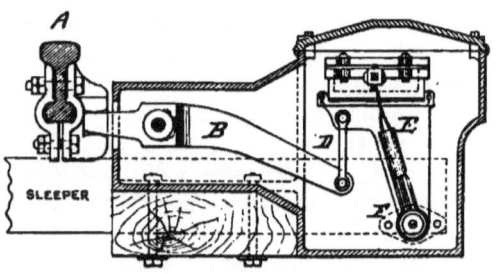

FIG. 92.

casting fixed to the signal post so that R always rises and falls in a parallel or vertical direction. Pivoted at m upon the rod R is a scale-beam lever M, one end of which is connected by a rod to the balance lever S at the foot of the post ; the other end being connected by a short rod to the lever L. L is pivoted at P on the main framing, and being practically in a vertical position will, when retained in that position by electrical aid, as will be explained further on, prove capable of forming a fulcrum upon which the lever M can act. When the balance lever S is worked by the pull of the wire at s, the rod R is lifted and the signal arm lowered.

The lever L carries a spring armature n, which, when the lever is in the vertical position, is brought within the influence of the poles of an electro-magnet fixed in a box E placed on the shelf F on the main casting. When the current is flowing through these coils, the armature n on L is held fast, and, as previously explained, forms a sufficient support for the action of the balance lever S. Directly the current ceases to flow, however, L falls away from the poles of the electro-magnet, consequent upon the slight inclination it possesses in that

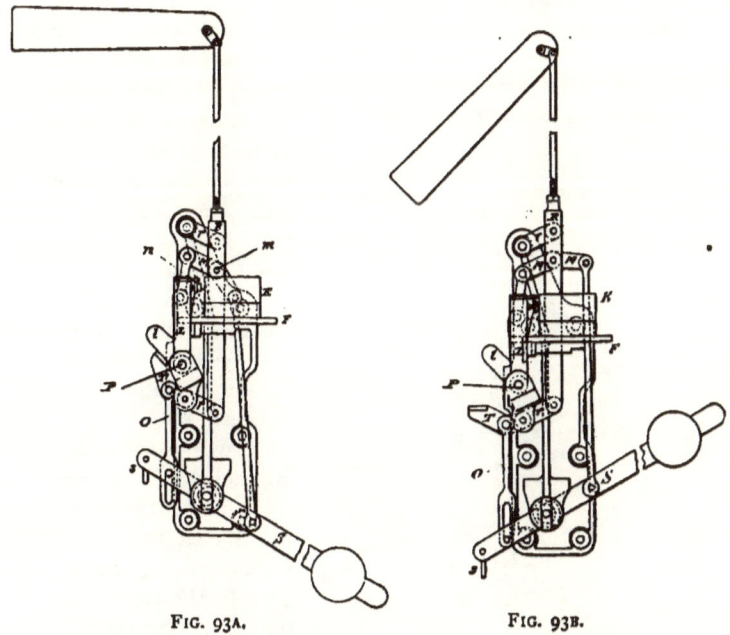

FIG. 93A.   FIG. 93B.

direction, and the absence of the influence of the armature n; and as L falls away it lowers M, allowing the signal rod R to descend and the arm to go to position C, Fig. 93.

It will be observed that the balance lever S, at its s end is connected to a slotted rod O which is connected to a crank T provided with a shoulder piece impinging, as seen in Fig. 93A, upon the limb l of the lever L. The object of this is that T may lift L, and retain it in the position which will bring the armature n into position with the electro-magnet poles in the box E, which possibly the freedom of the lever M might not sufficiently ensure. The electro-magnet in use on the slotted signal post is operated by a local battery brought into use by a relay fixed in the signal box, or elsewhere as may be desired. The spindle of the upper, or red arm, of the block instrument is provided with a platinum contact point which, when the arm is lowered, dips into a cup of mercury, and by that means connects the local battery to the electric slot coils, thus saving the battery action so long as the block signal is at danger. A very small amount of force is required to sustain L in its vertical position when the arm is being lowered, the pressure being nearly in a vertical line with L, and so falling upon its centre P.

It will be clear that in order to put this signal at danger the circuit with the electro-magnet E has to be broken. This is effected by the train operating a treadle arranged the necessary distance in advance of the signal. The signal is pulled off mechanically, as shown in Fig. 93B; the lever L having previously been lifted into the vertical position, is held by n so long as the current flows through E. On the train passing over the treadle the current is momentarily broken there, and L, no longer sustained, falls away and the signal rises to danger. The

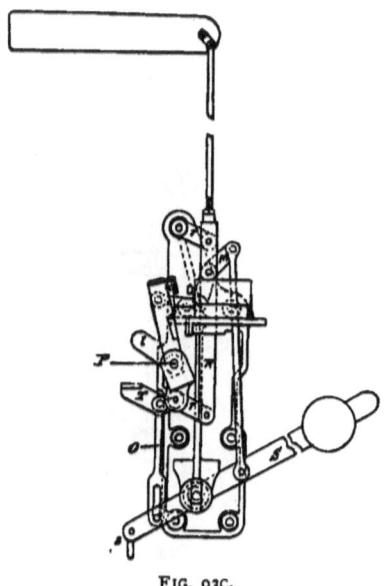

FIG. 93C.

signalman alone can pull the signal off: either the signalman or the train can put it on.

Messrs. Saxby and Farmer have also in operation a system of actuating points and signals by hydraulic means, the main object of which is the saving of manual labour and security in working. The system requires the aid of a pump, receivers, &c., in the same manner as the compressed air system in operation in America, and which has already been referred to in these pages.

SYKES' INTERLOCKING BLOCK.

The latest electric interlocking carried out by Mr. Sykes is that on the Great Eastern Railway, and it differs from that established on other lines by the addition of what is known as the *Train on* gear, by means of which the signalman has notified to him the signal last given. This is, moreover, so arranged in connection with the lever of the signal controlling the entrance to the section, that if the signalman in advance should omit to operate the indication "Train on" for his own guidance, and the "line blocked" indication for the guidance of the signalman in the rear, the signalman there will do so by the act of putting back his lever.

Fig. 94 is an external view of the front of the block instrument. Fig. 95 is a front view of the interior of the instrument; Fig. 96 a side view, showing the "train on" apparatus; and Fig. 97 is a side view, showing the switch at the back of the instrument, and which switches the locking instrument from the line circuit to the rail contact circuit, so that a current from the station in advance cannot take the *back lock* off to enable the signal to be put to danger, nor can a current from the rail contact circuit release the front lock to enable the signal to be taken off.

The principle of this system renders it impossible, when a train has been accepted from a signal box in the rear, for another train to be accepted until the first train has reached and passed the signal box from which the acceptance signal was given.

A positive current alone will unlock the lever, and when the signal is pulled off the same lock again locks the signal in the "off" position, so that the lever cannot be put back until the train for which the signal has been lowered has passed over a *rail contact* placed at a suitable distance in

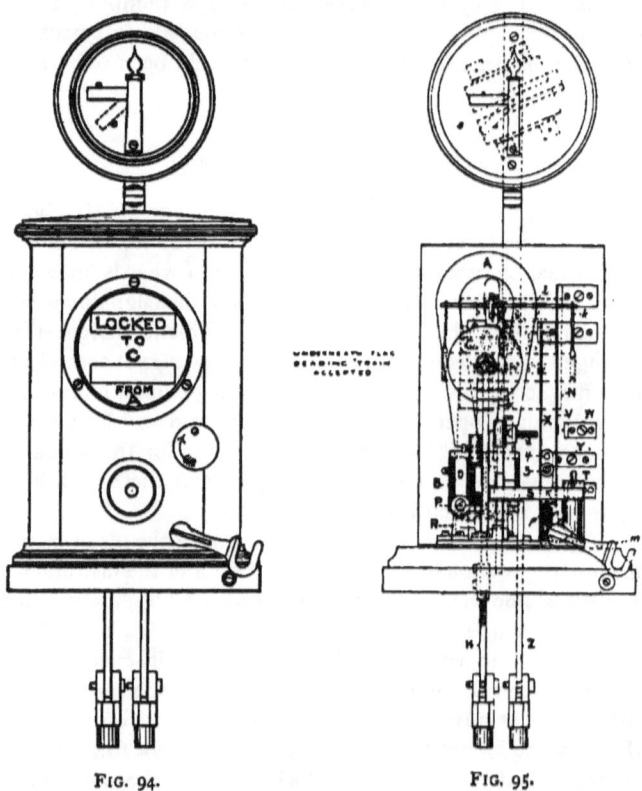

FIG. 94.   FIG. 95.

advance of the signal itself. The operation of this rail contact by the passing train removes the lock, and the signal can be again put to danger. It is then again locked, and can only be released after the train has passed clear out of the section in advance, and permission given for a following train.

In Fig. 95, A is a four-bar compound permanent magnet, on the soft iron pole pieces of which coils B are fixed. The armature C, Fig. 96, is attached to an angular lever E centered at D, on the upper extremity of which is a small wheel or roller F, which, when in the position indicated in Fig. 96, is supporting the rod H by means of the angle piece G. Attached to the rod H is another angle piece I, which carries a lifting piece K influenced by a strong spiral spring L. Pressing against the angle piece E is an adjustable extension spring O, the pressure of which is regulated by the adjusting screw P. The object of this spring O is to aid the current in discharging the armature C, and thus remove the wheel F from below the angle piece G. The tension of this spring must be such that the current will not discharge the armature without the aid of the spring, and that the spring shall not discharge the armature without the assistance of the current.

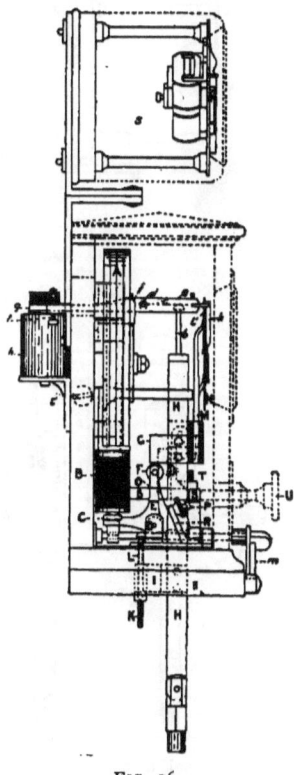

FIG. 96.

In accepting a train from the signal box in the rear the plunger U is pressed, and this action operates the angle piece S, the axis of which is at T. The spring V is placed in contact with W, and the battery connected to line. At the same time the spring X follows the spring V, and makes contact with Y. Immediately the plunger U has been pressed, the rod b, Fig. 95, carrying the indication "train accepted," falls, and prevents the signalman again using his plunger until this rod has been raised by taking off the signal lever and again putting it to danger.

When a train has thus been accepted and the signal "train entering section" received from the station in the rear, the signalman in advance turns the hook m so that it shall engage with the plunger. This action of m, by means of the cam n and the insulated piece p, breaks down the circuit to the instrument in the rear, causing the electric semaphore arm s there to rise to the danger position, and exhibiting on his own instrument the instruction "train on."

Should the signalman in advance have omitted to turn the hook m, the signalman at the station in the rear, on putting his own signal lever back after the passage of the train, momentarily breaks down the circuit by causing the knuckle joint w, Fig. 97, to depress the lever contact spring x, thereby indicating "train on" to the signalman in advance.

In order to prevent a train being accepted when it should not be, a device is provided on the rod H which prevents the use of the plunger to accept a train from the rear, unless the signal is not only at danger but locked in that position. In Fig. 97 the rod u is shown; this rod carries certain insulated plates which are capable of being, by the movement of the rod, placed in circuit with the various springs v, v^i, v^{ii}, v^{iii} and v^{iv}. By this means, when the

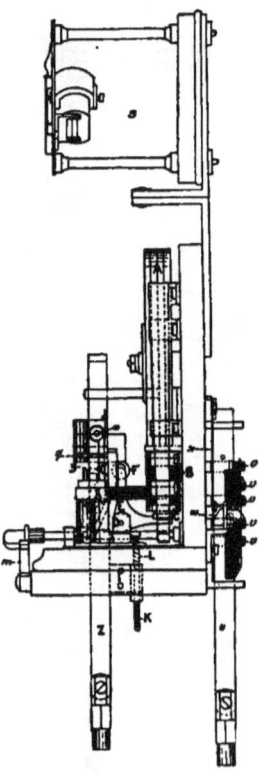

FIG. 97.

signal is at danger the line is joined through to the locking coils, semaphore arm and earth; but when the signal is off the line is put direct to earth, through the semaphore arm, and the locking coils connected to the rail contact circuit, so that the passage of the train may take the back lock off.

The "train on," and semaphore arm are maintained in position by an independent battery of two cells. The danger signal is produced by gravity.

The lock on the signal lever is effected by means of the rod H.

Now let us follow the several actions consequent on the passing of a train.

Assume three sections A, B, C, and that we are standing in the centre box B. The bell announces a train from A. When the necessary bell signals have been given and answered, B presses in the plunger of his instrument and the words "train accepted" appear on the dial of his block instrument. A now gives to B the "train entering section" signal, and B then turns his switch hook, which passes over the plunger, acting as a reminder that a signal has been sent, and causing the words "train on," to appear in the place of the previous indication "train accepted."

The train is now approaching B. The signalman at B signals its approach to C, and C on pressing his plunger unlocks B's starting signal lever. B's block instrument now reads "free" in the upper opening, and "train accepted" in the lower, the semaphore at the top being deflected.

B now being free to lower his starting signal, operates the lever, and the word "free" is replaced by "locked."

The locking mechanism of the instrument is now disconnected from the line wire in connection with C, and switched on to the rail contact, which is fixed a train length in advance of B's starting signal. B now gives the "train entering section" signal to C, and C turns his switch hook, causing the semaphore arm at B to assume the *danger* position. The train having passed out of B's section and depressed the rail contact there, the back lock is taken out of B's starting signal lever—the upper opening in his block instrument reading " Free "—and the signalman is at liberty to put it back in the frame and accept another train from A.

In replacing the lever in its normal position, the upper opening shows the word "locked" and the lower one exhibits no instruction whatever—being vacant or blank. The switch hook is taken off the plunger, indicating to A that the line is clear; and, similarly, when the train has passed C, B's semaphore assumes the clear position.

SYKES' ELECTRIC SLOT SIGNAL.

Mr. Sykes' "Electric Slot or Signal Arm Replacer" is shown in Fig. 98. Its object is to automatically place the signal arm in the danger position independently of the signalman by the passing of the train, the presence of which it is desired to protect. The "slot" apparatus is fixed between the balance-weight lever and the signal arm, and moves with the upright rod when the arm is lowered. The battery power required is small, and the current is in operation only during the passing of a train over the treadle.

M, Fig. 98, is a solenoid, of which A is the movable core. On A being, by the influence of the current passing through the coil M, drawn in, the hammer piece G is released, and in falling impinges upon the rod C, which is centered upon the extremity of the lever D, and which, it will be observed, is counterbalanced so that its normal position may be that shown in the figure. D is centered at d, and on C being struck by G, the force of the blow is sufficient to drive C down to such an extent as to release the roller arm E, which it will be seen engages with D at e. F, which is connected to the balance-weight lever, is a slide bar working within a slot, and having its end behind E inclined or bevelled. Ordinarily its influence is to press E against D at the engaging point e, and when E is knocked away, to slide up behind E and to force it into the dotted position.

The roller arm E is provided with a projection which,

when E is forced into the dotted position, rides upon the lower shoulder of G, until it, G, is again deposited upon B.

We will now follow the action of the several parts on a train passing over the treadle.

The signal is first set in the position for the reception of a train by the signalman pulling over his lever. This draws over the balance weight bar to which F is attached. F is raised, and, lifting the whole apparatus upwards, deflects the signal arm.

The train now passes over the treadle. A current is passed through M; A is attracted; the hammer G released; the pin C forced down; E is disengaged from F and carried into the dotted position. There is now nothing to sustain the apparatus, and the signal arm, by gravity, rises to the danger position. As when the signal arm was lowered by forcing F upwards, so, when the signal arm is raised to the danger position, is the box containing the discharging gear lowered. E is free, and as there is nothing to support the box it drops to the extent of the space between the top of the slide

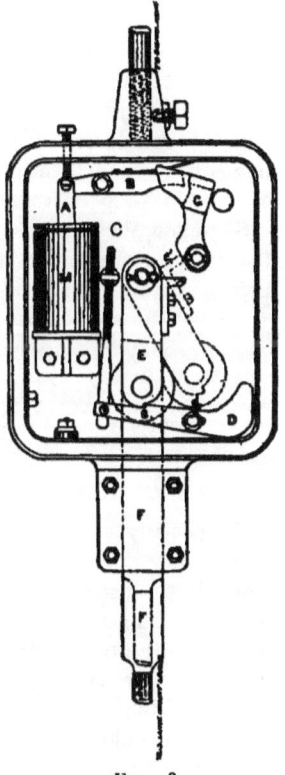

FIG. 98.

bar F and the point where it conflicts with E—a space sufficient to admit of the signal arm going fully to the danger position.

To reset the signal for a train to pass it, the signalman has to first place his signal lever ON. This will draw down F. E will follow it until it again engages with D at *e*. When in this position, on the signalman drawing over his lever, all parts being again locked, the balance-weight bar will force F upwards, and the whole apparatus will again be raised, and the signal arm lowered.

Mr. Sykes employs a double jointed treadle, as seen in Fig. 99.

Three iron brackets are secured to the rail, and the treadle is bolted to the two outer ones at B, B. The centre bracket carries an adjustable pin P, which rests upon the pivoted ends of two levers L, L, the extreme ends of which are immediately underneath mercury cups C, C. On an engine or coach passing over the spot P, the depression of the rail acting upon the extreme ends of L causes them to assume the dotted position, tilts the mercury cups, and thus makes or breaks contact as may be desired.

It will be seen that electrically controlled signals of this character may be employed in many ways to great advantage,

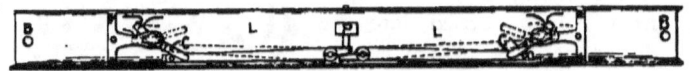

FIG. 99.

and especially so, perhaps, for siding working. It would also appear they might be of service for distant or other signals, worked at a distance by wire capable of being influenced to an appreciable extent by changes of temperature, which not unfrequently prevent men from either failing to pull off their signals properly, or to place them fully to danger. A passing train would in all cases put the signal to danger, and if the signalman could not get it off again without letting out his wire, he would be obliged to do so. Clearly the *danger* signal would, assuming of course the good working of the apparatus, be ensured, which is the most important point. Signals so fitted can be, and of course are, used in connection with the interlocking of the electric with the mechanical block signals.

All signals so operated should be electrically repeated, so that the signalman may be assured of their effectual working.

SPAGNOLETTI'S ELECTRIC LOCKING SYSTEM.

Spagnoletti's present system of electric locking consists essentially of the following parts for each road :—

1. A treadle.
2. A relay.   } At the station
3. A releasing instrument. } in advance.

4. A locking instrument. { At the departure station.

The operation of the apparatus is as follows. Assuming A and B as two stations ; that a train is required to travel from B to A ; that the line is clear, and that the instruments are in the normal position, ready to receive a train :—

B asks A to accept a train.

A depresses the key of his releasing instrument, which discharges the lock on the signal lever at B.

B then pulls over his lever and lowers his starting signal, which action on the part of B releases the armature of the relay at A.

The train moves out of the station, and B signals its departure to A.

B replaces his lever behind the train and it again becomes locked.

The train now arrives at A, and in leaving that station passes over a treadle, the depression of which resets the locking instrument at A, and A is then in a position to accept another train from B.

The operation of the different parts may be gathered from the accompanying diagram, Fig. 100.

D is the releasing instrument fixed at station A. D' represents the condition of the instrument when in the opposite position. The normal condition is that shown at D.

When it is required to bring a train from B to A, the key K of the instrument D is depressed. This raises the bar b to the position b', where the end of it engages with the catch c of the armature a, and is thus held in the horizontal position. The insulated pin p on the bar b is withdrawn from between the springs s, s, which are thus placed in contact, and the instrument now stands in the position D'. The depression of the key K also temporarily raises the line spring $l\,s$ to the contact e, and thus puts the relay circuit to earth. The relay R is then operated by the battery B, the circuit being from earth E^3, spring $l\,s$, contact e, battery B, relay R, and line wire to earth at station B. This attracts the armature of the relay R, and establishes connection at d. A permanent current then flows from earth E^4, d, B, R, and line wire to the lock L at

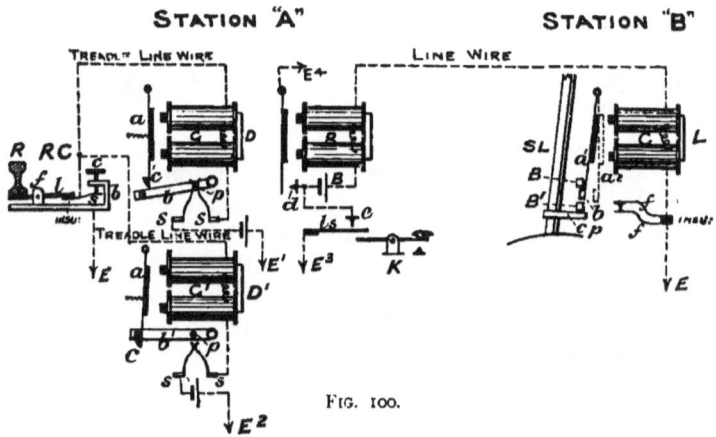

Fig. 100.

station B, passing through the electro-magnet C and the springs f, f to earth. The armature a' of the lock L is attracted and the lock released.

The signal lever is locked in the following manner. S L represents the signal lever; C P a catch piece fixed thereon; B a rigid block, against which the end b of the armature bears when in the position a'. B' is a block, capable of being moved vertically when the armature is in the dotted position a^2. The normal position of the lock armature is a', and it is then impossible to move the signal lever S L, because the catch piece C P cannot be raised on account of the sliding block B' being stopped by the block b' and B. When a current is sent from station A through the coils of the lock, the armature is brought into the dotted position a^2, and the block being thus removed, the catch piece C P and the block B' can then be raised and the signal lever drawn over.

The action of drawing over the signal lever severs the two springs f,f, which are otherwise in circuit, and the line being thus severed, the armature of the relay at station A falls away from the contact d.

When the train leaves B for A, the signalman at B replaces his lever and the armature of the lock falls, by tension, to the position a', thereby again locking the signal lever. On the departure of the train from station A, it passes over the treadle R C, placed some train's length in advance of the starting signal. This treadle consists of a lever l having its fulcrum at f; a spring s attached to the lever l, but insulated from it; a bridge b carrying a contact screw c fixed to the iron base of the treadle, and thus connected to "earth." Upon the train passing over the rail, the short end of the lever is depressed, thus bringing the spring s into contact with the screw c. The releasing circuit is now complete, and a current flows, for the time being, through the coils G'; the armature is attracted, and the bar b' released; the springs s, s are again severed, and the instrument is now in its normal position as represented in D, ready for being actuated by another train.

TYER'S INTERLOCKING BLOCK.

This instrument, Fig. 101, consists of two semaphores, a commutator E with bell plunger F; a screen G, and two additional screens H, I, working in combination with one another. The treadle in connection with the rails may be fixed at any point convenient for the purpose.

On turning the commutator E from right to left, the screen G is replaced by the "line clear" screen I; the lower arm at the sending station, and the upper arm at the station receiving the signal, is lowered, and the lock is removed from the starting signal. On the departure of the train, say from station A, the starting signal is again placed at danger by the signalman, and the station (B) to which the train is travelling advised by bell in the usual way. B now reverses his commutator, thereby mechanically replacing the instruction "line clear" by "train on line" in the rectangular space above the commutator, and at the same time raising the lower arm of his instrument and the upper arm of that at A to the danger position.

As soon as the train has passed over the treadle at B, the signal "train out of section" (G) is automatically caused to

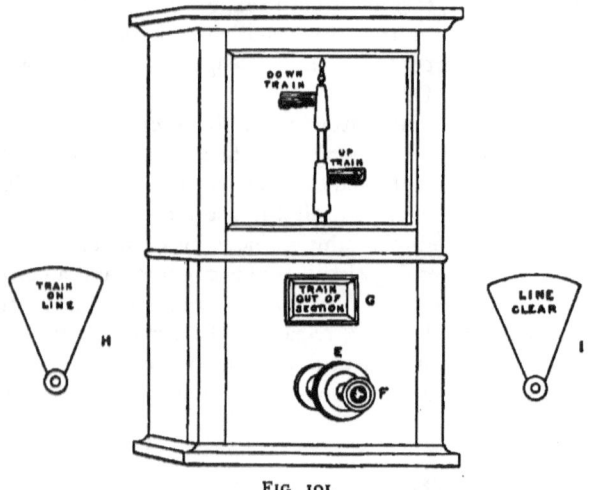

FIG. 101.

appear, as seen in the figure, on B's instrument, and the instruments at A and B for the road in question are again ready to deal with a following train.

The semaphore arms are operated in the usual manner by the combined action of permanent and electro-magnets. The commutator action may be ascertained by studying Fig. 102.

S^1, S^2, S^3, Fig. 102, are the three screens, H, I, G, respectively, mounted on the axle of the commutator E, Fig. 101. The screen S^2 (line clear) is *fixed* to the axle of the commutator E, and moves with it either to the right or to the left, as may be required.

The screen S^3—train out of section—turns freely on the axle, and is moved toward the left whenever the commutator is turned in that direction. This screen has a counterweight, the purpose of which is to restore it to its normal position when not otherwise influenced.

Whenever the commutator E is turned from right to left, the screen S^3 is moved on one side, leaving the signal "line clear" on screen S^2 visible.

Screen S^3 has also a projecting pin P fixed at its lower edge, and whenever the screen is moved this pin passes the

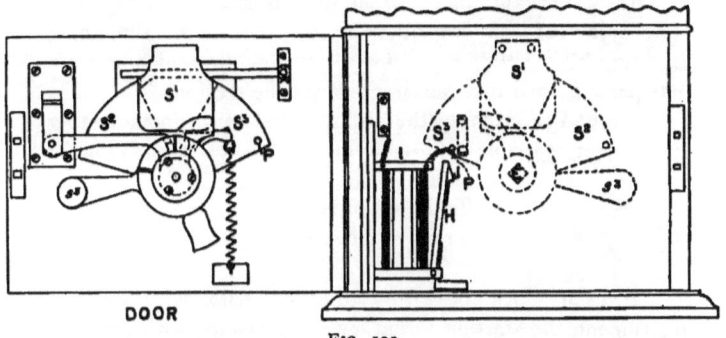

Fig. 102.

rib J attached to the upper portion of the armature H of the electro-magnet I, and is there retained.

On the receipt of the departure signal, the commutator E is turned from left to right, disclosing the signal "train on line."

On the arrival of the train at B, and on its passing over the treadle there, the local circuit in connection with the electro-magnet I is completed. I becomes energised; H is attracted; the pin P of the screen S^3 released, and the screen, under the influence of the counterweight, rises, covering the signal "train on line," and exposing that of "train out of section."

When once the signal "line clear" has been given and

acknowledged, it cannot be transmitted a second time. The train must pass through the section and over the treadle at the incoming station.

The apparatus can be so arranged that by the aid of a treadle at the departing station, on the train passing over the treadle it shall put the starting signal to danger, and at the same time automatically raise the electric semaphore arm for the road on which the train is travelling.

TYERS' AUTOMATIC SIGNAL LOCK.

Mr. Tyers' automatic signal lock is shown in Fig. 103. The object of this invention is to prevent the signalman lowering his starting signal for a section into which a train has passed, until the train has cleared the section.

The end of each section is provided with an indicating instrument capable of rendering either of the following signals :—

"Train out of section."
"Train in section."

At each block post, in connection with its respective instrument, the starting signal lever is provided with a lock as shown in the figure, which locks the lever in the position shown.

Connected with the railway metals, rail contacts are fixed —one, that at which the train enters the section, for locking the lever; the other, at the distant end—where the train is clear of the section—for releasing the lock. It will be obvious that the same contact arrangement is capable of locking the lever of the section which the train is about to enter, and at the same time releasing that for the section in the rear.

One line wire only is employed to work the signals in both directions, up and down. It will thus be understood that the signals rendered are of a momentary character.

The same description of lock is intended for use when shunting operations are being carried on Locking and releasing rail contacts are arranged as required. A down goods train has arrived at a station. In approaching it has operated the contact, which automatically causes the signal "Vehicles on line" to be presented on the indicating instrument, and at the same time locks such of the down signals as may be required.

If the train simply passes through the station, it, on passing out at the other end, passes over another rail contact which releases the lock.

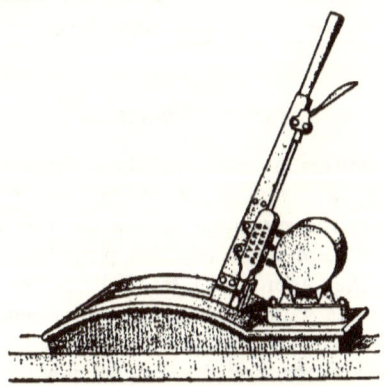

FIG. 103.

If, however, it has to shunt, the releasing signal is not given until the train has passed into the siding and operated a releasing contact there.

When the train comes out of the siding it again passes over the locking contact, and the signal lever is again locked until released by the train passing over the distant contact on the main road at the far end of the station, when the instrument records "Vehicles clear of down line."

CHAPTER IX.

MISCELLANEOUS APPLIANCES IN CONNECTION WITH BLOCK WORKING.

SIGNAL REPEATERS.

ELECTRIC repeaters for recording the position of the signals to which they are applied, have become practically an integral part of the signalling system of every railway. So necessary have these useful adjuncts proved, and to so great an extent are they appreciated, especially at heavily signalled termini, that it is with difficulty a position suitable for the purpose for which they have been instituted, i.e. for bringing prominently under the eye of the signalman the condition of the signal operated by him, is found within the region of the signal frame. The importance of these little automatons is second only to that of the signal which they represent. This is shown in the numbers which are now to be found in use on every important railway.

In their application, the contact maker in connection with the signal should be so arranged that it may be actuated direct from the signal arm or the signal itself, whatever may be its form. Any intermediary connection is undesirable; with every such intermediary there is increased liability of failure, therefore the connection with the signal should be as direct as is possible.

The repetition extends in general to three positions only, viz. to the signal ON, signal OFF, and that position which is neither "on" nor "off" but intermediate between the two.

That position which determines when a signal is ON, and

when it is *not* ON, is a very important one, and often leads to very grave questions. It is therefore most desirable that the contact maker should admit of just that record of the ON position which is recognised as the *danger* signal and *no more*. That is, that the instant the signal assumes the position of *danger*, the repeater shall show that the signal is ON, and the instant it deviates from this position the repeater shall indicate that it is *not* ON.

This is easily assured. The signal itself works within certain limitations: it can be pulled "OFF" to a certain extent; it can be put "ON" to a certain extent. There are stops to both positions. The ON position may be regarded as the movement of the arm within an arc of three degrees from the horizontal in a downward direction. The ON contact of the contact maker should provide for this, and to counteract any deviation therefrom by the lineman, this contact should be a fixed one. The OFF contact may be adjustable, in order to meet the drop of the arm, whether falling within the sheath of the signal post or merely to a caution position.

The wire operating the signal, it is perhaps needless to say, is also a governing factor, not in all cases for good. Subject to conditions of temperature, this wire expands and contracts, and thus, unless the signalman is assiduously attentive to these changes, the action of the signal is impaired. If the wire is contracted, the arm cannot go to "danger" when desired, and if the wire has become too long from expansion under an increased temperature, the signal cannot be pulled off to the "clear" position. It is in the presence of such variations as these that the electric repeater acts as an invaluable check. Where signals are obscured from the view of the signalman by curvature of the line, by structures, trees or other impediments, the repeater is indispensable.

If the batteries are placed at the signal post, one line wire only will be required.

The contact maker employed on the Midland is shown in Figs. 104, 105. A is a galvanised cast-iron box, capable of being fixed in position by screws as shown, carrying within the circular portion the moving part, which, acting in unison with the signal arm, effects the electrical arrangement necessary for rendering the repetition. D is a slotted lever, fixed upon the centre pin F, which latter has fixed to its extremity within the box A, a tongue I capable of passing between the springs G, the ON contact, and H, the OFF contact. Each of these contact pieces is provided with springs, so that the tongue piece I may, in passing between them, ensure perfect contact. G is so arranged that it provides for a movement of the arm of 3°; the contact-maker has therefore to be fixed on the signal post in that position which will afford this record.

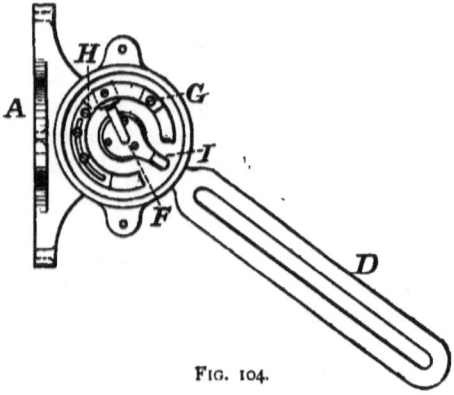

FIG. 104.

H is slotted so as to admit of adjustment under two binding screws arranged in the path of the slot.

Fig. 105 will explain the position in which the contact maker is required to be fixed in relation to the signal arm B. A is the cast-iron box containing the contact parts, D the slotted lever, and E a pin firmly fixed to the signal blade by the plate E'. The arm works on the centre C, the slotted lever D works on the centre A. They thus form different arcs, and hence the need for the slot in lever D. If the movement of the signal blade is followed, it will be seen that when it has passed through some 40° the pin E will be at the extreme end of the slot. If now we turn to Fig. 106, the action of the contact maker and the electrical arrangement will be readily understood.

The repeater instrument generally employed on the Midland Railway is shown in the accompanying cut, Fig. 107. It is, in fact, a single-needle coil. The needle moving inside

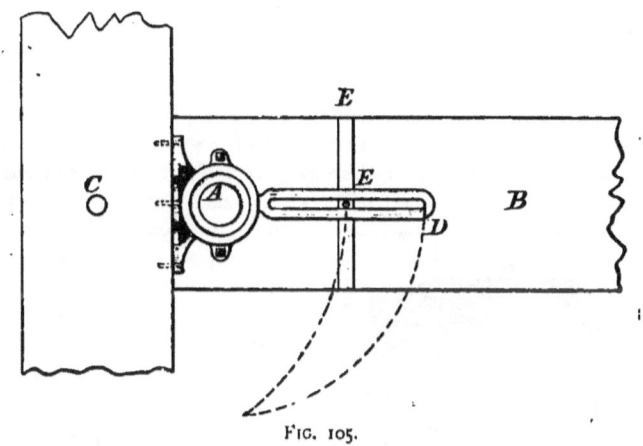

Fig. 105.

the dial, carries a flag inscribed OFF, ON. Either indication is brought up to the aperture according to the direction of the current, causing the indication to read "Signal OFF," or

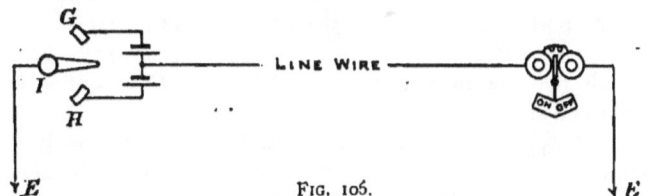

Fig. 106.

"Signal ON," as the case may be. Should the tongue of the contact maker fail to make contact, the flag hangs midway, affording a mixed indication, neither ON nor OFF.

These instruments are made to occupy the space between any two levers centre to centre, in practice 6", so that they may be fixed on the face of the instrument shelf over the lever working the signal, the action of which it is required to repeat.

There are various forms of repeater instruments, some representing the signal itself—a miniature of the signal—others with inscriptions upon the indicator ; and again, others having an indicating needle only—as in the single-needle block—with the signals which the movement of the needle is intended to indicate inscribed upon the dial. It is obvious the record may take various shapes, and that the instrument itself may be made of such form as will best meet the position in which it has to be placed, so long as its indications are clear and decisive and sufficiently open to observation.

FIG. 107.

Inasmuch as in the repetition of a signal the danger lies in an indication that the signal is "on" when it is *not* in that position, or when it is in a position which the driver of an approaching train would not consider to be the danger position, it is desirable the entire arrangement under which the repetition is obtained should be such that the failure of any portion may result in the production of an indication, either that the signal is "off," or that it is "not on." This entails that the current by which the "danger" repetition is produced shall be active when the signal is "on."

It has, in a previous paragraph, been pointed out how important it is that the contact under which this indication is produced should be so limited that it shall cease the instant the signal attains that position which is regarded as "*not on.*"

Signal repeaters are, in common with all electrical and other apparatus, liable to failure. A failure of the battery, the "earth," or the line wire, should result in the indication being of a negative character. The indicator should show neither ON nor OFF.

If the signal itself fails to go to danger when required to do so, the repeater should afford a similar indication. Now, if the signalman should in such a case assume that the repeater is out of order, it is clear evil might result. Signal repeaters should be so reliable as to earn the complete confidence of the signalman. It is obvious that when a signal fails to assume the danger position when required to do so, the consequences may be serious. Therefore should a signalman when he places, as he assumes, his signal at danger, find that the repeater indicates it is not in that position, he should be instructed to regard the signal out of order and to take the usual steps for the protection of the traffic.

The contact maker indicated is but an example of that which is largely employed, not only by the Midland Railway but by others. It will be clear that diverse forms may be designed, and found, possibly, equally useful for the purpose. The chief desiderata are that it shall be reliable, capable of accurate adjustment, strong, easily and readily applied, and durable. These are the points aimed at in that described, and which the experience obtained from the large numbers in use induces the author to believe have been achieved. It will be found useful for many purposes other than that to which it has been made applicable in the foregoing remarks as, for instance, when the lever is replaced by a rod, terminated by a float, for recording the rise and fall of water in a tank in which the water is required to be maintained at a certain height.

REPEATER DISC.

This instrument is employed to indicate to the signalman when the controlling lever of a signal which requires the joint operation of himself and the signalman at the adjoining box, is "off" or "on." As employed on the Midland Railway, these instruments are fixed in a position on one side of the signal frame, somewhat the same as that occupied by the old form of mechanical "disc"; and in order to dissociate them from the signal repeaters, which they to some extent resemble, they are provided with a flat brass rim, as shown in Fig. 108, on which is inscribed the name of the signal.

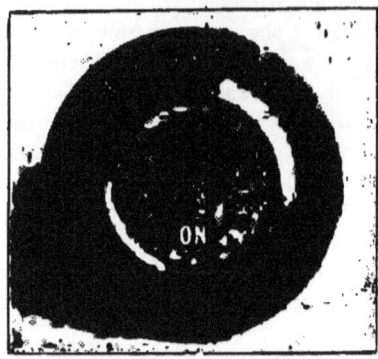

FIG. 108.

As these instruments are not always under the immediate observation of the signalman, and whereas if they were it is possible a change in the indication from ON to OFF, or *vice versâ*, might take place without being noticed, the reversal of the signal is accompanied by the ringing of a trembling bell fixed in its neighbourhood. This is effected by attaching to the spindle carrying the indicating flag, Fig. 109, an extension a, in such a manner that it shall, when the indicator moves from one side to the other, pass through a bath, b, of mercury. A trembling bell and a battery are in circuit, as shown in the figure. During the time the extension a is in contact with the mercury, the local circuit is complete and

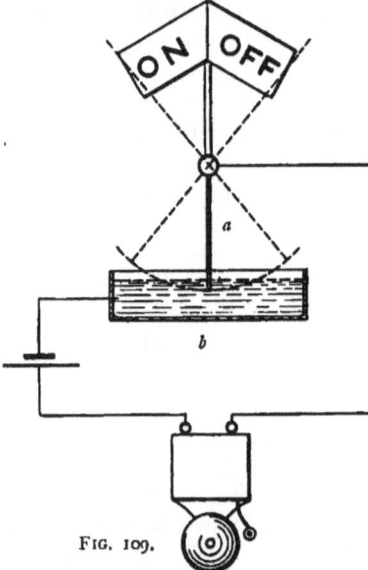

FIG. 109.

the bell will ring, and so call the attention of the signalman to the movement of the indicator. One bell will serve several such indicators.

LIGHT INDICATORS.

It is by the aid of artificial light that signalling is conducted during the absence of daylight. Without this artificial light no visual signals could be formed. By its aid, combined with the movement of the arm (to which is attached the *spectacles*), we obtain the necessary night signal. The existence of the light is therefore all-important, and hence the necessity for its record.

This is effected by arranging a form of *pyrometer* over the flame which forms the light; and if the adjustment of it is proper, it can be made to record by the aid of a suitably constructed instrument, not merely the existence of the light so long as it is a *good light*, but to forcibly call attention to any practical diminution or failure of the light, by starting, and keeping in action, a trembling bell until switched off by the signalman.

In some instances the pyrometer, or "*expansion bar*," as it is generally called by those who have to deal with it, is so arranged that on its expansion under the influence of the heat from the flame it shall *break* the electrical circuit, and the instrument shall, in the absence of the current, by gravity or a tension spring, be caused to indicate "light in." Such an arrangement is undesirable. A broken wire or a failure in the battery would occasion the production of the same signal while the light itself might be out!

The indication "light in" should be produced and maintained by the energy of the current, and any failure of battery or wires should result in the intimation that the *light is out*. It will then be the duty of the signalman to satisfy himself of the condition of his signal.

The form of expansion bar in use on the Midland is illustrated in Fig. 110. G is a metal tray on which the lamp is made to slide. The lamp itself is a covered oil lamp, as shown in Fig. 111, and, as will be seen, is provided with a hole for the reception of the expansion bar. Fitted permanently to G is an upright bar of iron H, which is slotted so as to receive the pins g, g' of the iron frame A of the expansion bar, and ensure its rigid attachment

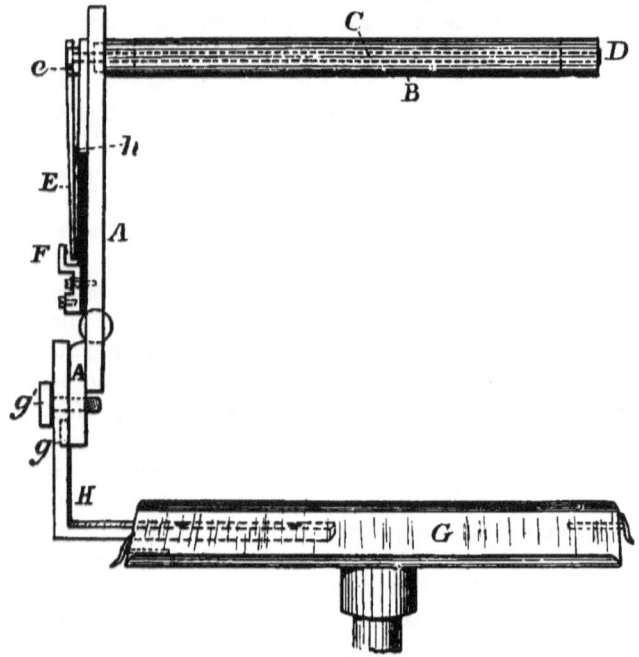

FIG. 110.

thereto. B is a brass tube formed of metal ·0313 inch in thickness, the outside diameter of which is ·5616, or $\frac{9}{16}$ of an inch. The end D is closed by a stout disc of metal, while its opposite end is permanently fixed to A. C is a steel wire of No. 10 L. W. G., fixed to the metal disc D, having its opposite end pressing against the lever E which is pivoted at e; F is an insulated cock to which the line wire is attached;

h is a flat steel spring, the purpose of which is to press the lever E against the line connection F when the expansion of B will admit of it. The frame A is to earth.

On being brought under the influence of the flame of the lamp, the tube B will expand and carry with it the steel rod C, the expansion, owing to B being rigidly fixed to A, being in

FIG. 111.

the direction of D. As C leaves the short end of the lever E the spring *h* will assert its influence, until E is pressed against the contact stud F, and the line wire thus placed in contact with earth. The circuit is now complete, and the indicator tells the signalman the light is IN.

Any further expansion of B will have no influence on E.

The adjustment of the bar is made in the workshop, and is such that there is little need for readjustment after the bar is fixed.

Now assume the light to go out or to become low. The tube B will be the first to feel the loss of its warmth. It will

FIG. 112A.

contract to a certain extent. The wire C will not have experienced the same loss of temperature, and it will remain extended almost to the same degree as when the light was burning. The consequence will be that its influence on E will be speedily felt, and the circuit between F and E will be broken.

to Railway Working.   201

We will now go to the instrument employed in conjunction with this expansion bar. It is shown in Fig. 112, and consists of a trembling bell and an indicator, the *latter* being mounted upon the spindle of a magnet working within a pair of coils. The indicator carries the words IN and OUT. The expression IN, causing the instrument to read $\genfrac{}{}{0pt}{}{\text{LIGHT}}{\text{IN}}$ ',

FIG. 112B.

is brought up to the rectangular opening in the screen under the influence of the current, which, as previously explained, is brought into action on the expansion of the bar arranged in position over the light. So long as the circuit is completed through the expansion bar a constant current flows and the indicator remains fixed ; but on the circuit being broken at the expansion bar, another, but one of an intermittent character, is opened by means of the armature and the make and break spring in connection therewith — see diagram, Fig. 113. The armature carries the bell hammer, and the arrangement being that of a trembling bell, so long as the circuit by way of the expansion bar is broken, the local circuit will actuate the armature, and a continuous ringing of the bell will be established. A switch is provided in the front of the instrument in order that the ringing, when once the signalman's attention has been drawn to it, or at such times as the light is not in operation, may be suspended.

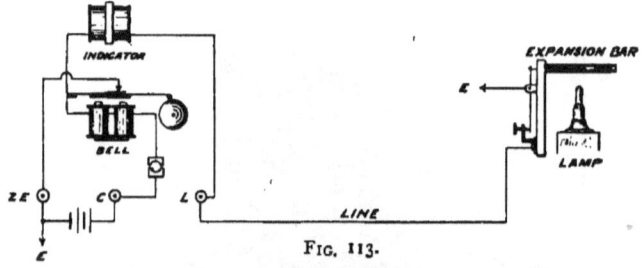

Fig. 113.

The signalman should, in order to test the efficient action of all such instruments, turn the switch to ON prior to the expansion bar being brought under the influence of the lamp.

As with repeaters, so with light indicator instruments, the form may be varied at will, as also the method of securing the signal. It is desirable, however, that in case of failure of the light the signal should be continuous, acoustically and visually, until the defect has been set right. An instrument which merely calls attention to a failure by one touch of the bell, is by no means so effectual as that which continues to call attention until attention is secured.

TRAIN INDICATOR.

The instrument designed by Mr. Tyer, and shown in outward form by Fig. 114, is exceedingly useful for announcing the description of train approaching junctions, busy

to Railway Working. 203

station yards, &c. It consists of two portions—a receiving instrument A and a transmitting instrument B. The receiving portion has a dial plate upon which are recorded the several signals or indications required to be transmitted.

The transmitting portion also comprises a dial affording corresponding indications, attached to which is a metal disc

Fig. 114.

C, pierced with holes to correspond with the several indications. A peg D for use in connection with these holes is affixed near.

Upon the right-hand side of the sending portion of the instrument is a sliding bar and button E, capable of being drawn out to the dotted position F.

The internal arrangement of the apparatus is illustrated

in Figs. 115, 116 and 117. Figs. 115 and 117 refer to the receiving portion. A A' is an electro-magnet of which B is the armature centered as shown, and carrying on its axis a pallet

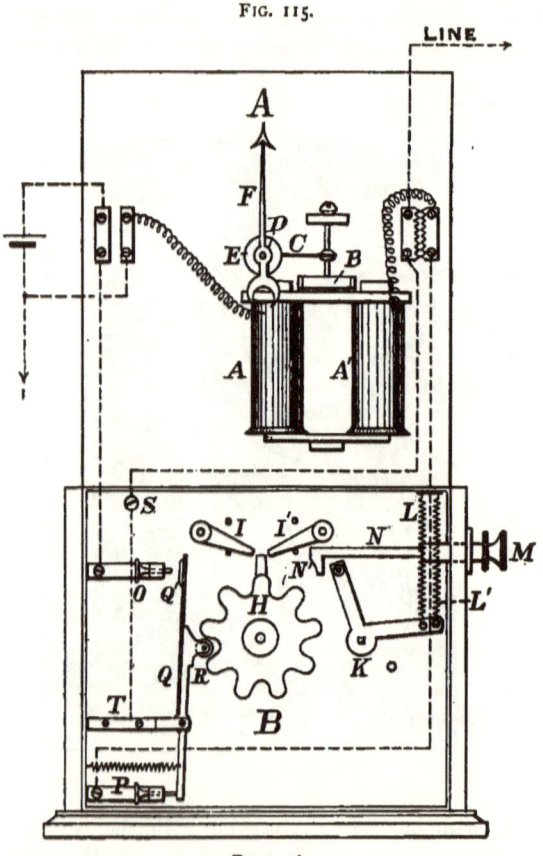

FIG. 115.

FIG. 116.

C ; D D' are two crown or toothed wheels mounted on an axle E, upon which is mounted the pointer F ; G is a helical spring, the purpose of which is to retain the pallet C against the wheel D'.

to Railway Working. 205

On a current passing through the coils A, the armature B is attracted towards the pole pieces of the cores of A, and the pallet C is then pressed against the wheel D. On the cessation of the current the helical spring G reasserts its influence, and C is then pressed against D'. It is always engaged with one or the other of the crown wheels, and is so arranged that when leaving the tooth of one wheel it shall engage with that of the opposite wheel. Thus, step by step, or tooth by tooth, the wheel is moved round—first one step by the current, then one step under the influence of the spring G.

Fig. 116 deals with the transmitting portion. H is a star wheel driven by a train and flier in the usual manner; I I' are two pawls retaining the wheel in the zero position; K a

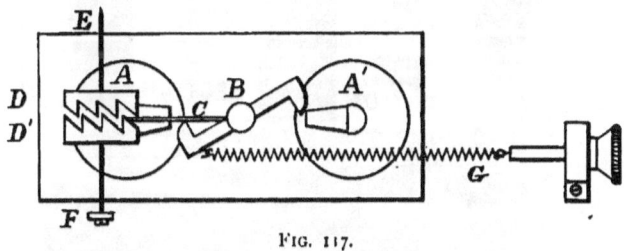

FIG. 117.

crank lever influenced by two helical springs L; O is an insulated contact point in connection with the + pole of the battery; P a similar contact point in connection with the electro-magnet A; Q is a rocking contact lever making contact with O or P, as influenced by the roller wheel R when moved in or out of the star wheel H as it revolves.

Whenever the button M is drawn out and allowed to return under the influence of the springs L, the pawl is released and H revolves until stopped by the pin D at the point where it is required to afford the necessary indication. Thus all the signalman has to do, assuming the propelling gear is wound up, is to insert the pin D in the hole opposite the indication required to be made, and pull out the button M.

SIGNAL INSTRUMENTS FOR LEVEL CROSSINGS.

At all points where roadways pass over railways on the level, it is essential a form of indicator capable of affording warning of the approach of trains to those in charge of the gates should be established. It is preferable this indication should be acoustic as well as visual.

FIG. 118.

Such an indicator is shown in Fig. 118. The visual signal is supplemented where requisite by a bell. The indicator instrument is inserted in the block circuits, one disc being employed for the up line and one for the down line. If the bell is placed in circuit on the bell wire, the gateman can hear all bell signals passing, and is able thereby to judge of the approach of a train. He also sees by the indicator instrument when a train is approaching and when it is IN the section; the former signal being produced by the current which, on the block instrument, produces the "line clear" signal; the train in section is produced by the "Train on line" signal.

Mr. George Lopes, of the London, Brighton and South Coast Railway, has recently devised an independent method of warning gatemen of the approach of trains. At the necessary distance from the crossing he arranges, in connection with the metals, a treadle commutator which, on being operated by a passing train, sends a current of a given polarity through one set of the coils of a differentially wound relay, which completes a local circuit and starts a bell ringing. The bell continues to ring until the train has passed over another treadle, placed a sufficient distance beyond the crossing to admit of the clearance thereof of the longest possible train, when a current is sent through the opposite winding of the coils, which breaks the local circuit previously formed by the relay, and restores the apparatus to its normal condition. This arrangement admits of one battery serving the two circuits, the battery being located at the same place as the instrument.

MOVABLE BRIDGES.

The arrangements for the protection of traffic over movable bridges should be of the most complete character. In its application these arrangements will necessarily differ in order to meet the various systems of block signalling, &c., but in principle they should be such as to ensure—

1. That it shall not be possible to open or break the road over the bridge without the consent of the signal boxes on either side of the bridge.

2. That to give this signal it shall necessitate the drawing over of a signal or other locking lever in the signal box; that this shall only be possible when all signals leading to the bridge are at danger, and the road to the bridge diverted so that no train can pass on to it; and that when the lever affording permission for opening the bridge is thus drawn

over it shall lock all others in that position which shall stop all traffic and divert the roads so that nothing can pass on to the bridge.

3. That the locking lever referred to in the preceding paragraph, and which might be termed the *bridge lock lever*, shall be electrically locked in that position in which the road over the bridge is maintained for the passage of a train, so long as a train is being signalled on the block instruments governing the section ; and

4. That the drawing over of the bridge lock lever from its position of rest shall so interrupt, or so arrange, the block apparatus that it shall not be possible to accept a train for either road comprised within the section in which the bridge is situate.

LIGHTNING PROTECTORS.

The desirability of providing all underground wires, as well as all classes of instruments, whether employed for block signalling or for telegraph purposes, with a suitable protection against the effects of serious atmospheric electrical discharges will be self-evident.

There are three forms of *lightning protectors* which may be regarded as in general use.

That formed by a pair of No. 35 copper wires, insulated with silk and passed through a bath of hot paraffin, twisted together, forms an exceedingly simple and efficient extemporary protection, but usually at the cost of breaking down the circuit. The two wires are inserted, one in the incoming, the other in the outgoing wire. If the twisted wires are wound around a metal reel, and the reel is connected with earth, the current will, on the destruction of the insulation of the wire by the passage of the atmospheric current, be carried to earth. This, of course, is not necessary at terminal instruments, as the connection with one of the wires will be *earth*.

The fact that the protection of the instrument is only secured by the interruption of the communication, renders the general employment of this form of protector undesirable.

to Railway Working.

The most serviceable form of protector for general use is that shown in Figs. 119 and 120. It is formed of brass plates, jagged as shown in the illustrations. The jagged portion should be slightly undercut, so as to afford a clearance from the possible accumulation of dust, and at the same time add to the effect of the points, which, it is perhaps needless to add, should be opposite one another.

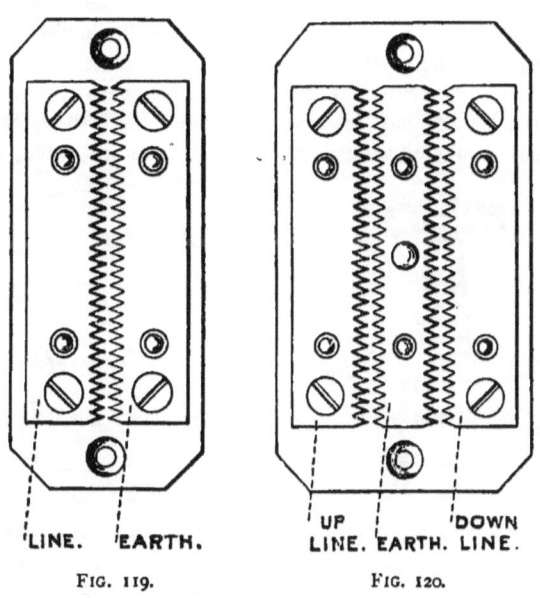

LINE.   EARTH.   UP   DOWN
                       LINE. EARTH. LINE.
FIG. 119.   FIG. 120.

Fig. 119 illustrates that suitable for apparatus which is usually in direct connection with the earth.

That shown in Fig. 120 is more suitable for intermediate instruments. The centre plate is connected by means of a terminal at the back, or other convenient part of the instrument, to earth. The current is thus induced to pass direct to earth at the first instrument it enters, instead of passing along the line wire and possibly through other instruments before doing so.

The *plate lightning protector*, Fig. 121, has not been so

generally used by railway companies. The lower plate B is connected to earth, the line wire is placed in connection with the plate A by means of the terminal attached thereto. The insulation between the plates is effected by the interposition of the plate of mica C, which is pierced in order to afford an opening or air space between the plates for the passage of any atmospheric electrical discharge. The mica employed for this purpose should be that known as Canadian amber mica, and of good quality. In thickness it should be from $4\frac{1}{2}$ to 5 mils, the outside edges turned and polished, and the holes carefully drilled so that the various layers of which the material is composed are not separated. With the mica well selected and conforming to the conditions indicated, these

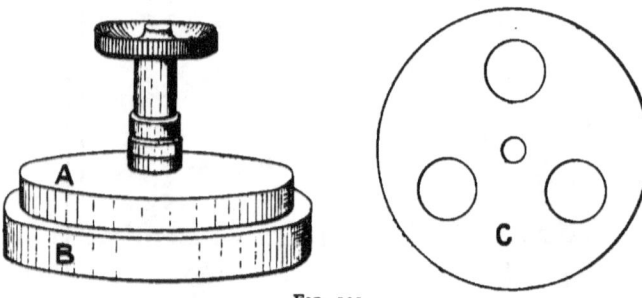

FIG. 121.

plates may be fairly relied upon to carry off electrical discharges likely to prove injurious to wires or instruments. It is necessary the plates should be occasionally wiped free from dust and any dampness that may gather upon them as the passage of a high tension electrical discharge will, in passing from the line to the earth plate, carbonise any particle of dust which may have gathered there, and thus place the line to partial earth. If dampness is present it will, in a like manner, allow a portion of the working current to pass to earth. It is a good plan to examine this form of protector every month, and in addition after every severe atmospheric electrical disturbance.

A still more sensitive protector is that shown in full size

by Fig. 122, and which is composed of a small vacuum bulb having a platinum wire A and B imbedded in either end of the tube, as represented in the figure.

These *vacuum tube protectors* are in general use by the British Postal Telegraph Department, at whose instance they were designed, and regulations for their periodical verification drawn up. The value of a vacuum conductor presides in its power to pass such currents as would prove prejudicial, to earth, while yet not interfering with such potentials as are necessary for ordinary telegraph work. In this the degree of rarefaction imparted to the air, together with the space appointed between the ends of the wires within the tube, are important factors. The condition of the tube is ascertained by passing through it currents of a given potential, derived from a small induction coil designed for the purpose.

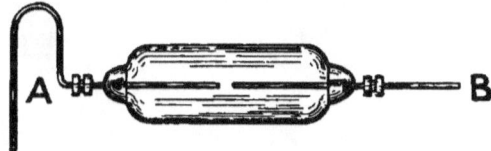

FIG. 122.

A six-block agglomerate cell is used with the coil, and in order to prove the value of the battery in combination with the coil, it is necessary that a $\frac{1}{8}$-inch spark, but no more, should be obtained from the secondary coil on the battery circuit being completed through the primary.

If, under these conditions, a tube is placed in circuit with the secondary coil, and a current from the battery cell indicated is passed through the "make and break" contact of the primary, a *glow* will be observed in the vacuum tube, provided it is in good condition. If the glow is absent the vacuum is defective, and the tube useless for the purpose required. It is important the potential of the battery employed should not be increased. A higher battery power would enable apparently a satisfactory result to be obtained when testing the tubes, while the value of the vacuum might have become so reduced as to largely increase the resistance of the air space between the line and earth connections, with the consequence that the discharge which it was desired should pass to earth would be allowed to pass into the cable or wire.

There is no reason why more than one form of, or more than one protector should not be used in connection with important wires. They would, of course, be connected in parallel.

CHAPTER X.

ELECTRIC LIGHT AND POWER.

THOSE who are acquainted with the railway working of to-day may possibly be able to form some idea of what would be the condition of traffic conducted without the aid of electricity. Certain it is that railway traffic would have to be largely curtailed, and it may also be averred that the absence of the present control exercised by the electric telegraph would tend to an increase in casualties. Of its utility and its numerous advantages there can be no question. Great, however, as has been its utility, there is little doubt that it is destined to be of immeasurably greater service, and that at no distant date. In the electric lighting of our railway stations and yards, and as an agent for power purposes, its advantages are already becoming more and more apparent. At all large railway centres constant demands exist for light and power. Light and power are the provinces of electrical energy. Electricity possesses the advantage over most other sources of energy that it may be generated at any convenient spot and conducted, by means of cable or wire, to any point at which it may be required for use. The loss in its transference thus from point to point is determined by the size of the conductor, and becomes purely a question of interest on outlay and cost of upkeep *versus* cost of production of the energy consumed. There is no condensation, no freezing. A temperature below freezing aids rather than impedes its power. Its distribution admits of more accurate determination than is possible with steam, hydraulic, or pneumatic services. That it should—allowing, for the present, the question of locomotion to stand on one side—prove the power of the future for all work now carried on by means of isolated plants, few will doubt. Economy presides in the centralisation of all such sources of power, whether applied to lighting or motor work. In centralisation we save not merely staff expenses—a very material item—but, in the application of suitable units of power, much initial cost in fuel and labour. Add to this the cost of buildings and the cost of land consequent upon isolated plants, and we have before us the main reasons why central electrical generating plants should command that attention to which the advanced ideas of the period entitle them.

Some of the advantages of the electric light in its application to railway work have, only recently, formed the subject of a paper read before the Institution of Electrical Engineers. In this paper it has been pointed out that although most of the larger railways have employed the electric light for some years, and many of them prior to its introduction on the Midland, the latter has far exceeded all others in the extent to which it has been applied, and it is upon the works carried out and the experience gained by that company that the paper referred to was based.

Incandescent lighting has in relation to railways so far found its chief employment in the illumination of offices, waiting rooms, and other spaces of limited area. In offices where a staff of workers are in constant attendance, or where, from the exigencies of the moment, work has to be carried on to a late hour of the night, its value in a hygienic sense cannot be overestimated.

The argument against the electric light for the illumination of offices, waiting rooms, &c., is usually one of expense. With a small installation the cost is, at present, in most instances greater than that of gas; but it is open to doubt whether such is really the case when the cleaning and redecorating of such places is taken into consideration. If we save by the electric light but one cleaning in three, it is probable that even the drawbacks of a small installation are fully met.

If to these arguments we add that of health, we have a

factor which carries with it far greater weight, and merits far more consideration by us; for health is money in more ways than one. It is more than money to the employed. To the employer it means more constant attendance, greater energy, and greater devotion to work on the part of the employé.

Whether the incandescent light may not yet prove of considerable service for upper floors of goods warehouses, where light is only occasionally required, is purely a question of the load factor. With a good output the cost will well compete with gas; and there are, in fact, very few railway installations which will not compete with gas, enormously in favour of the electric illuminant.

The directors of the Midland Railway Company have only recently installed the incandescent electric light throughout the offices, waiting rooms, &c., of the Derby station and headquarters offices. There is but one opinion of the result. The staff already feel the benefit of it in point of heathfulness and comfort, and the rooms and offices are practically as clean at the present moment as when the light was started. Although in this instance the demand for light is an exceptional one, due to a great mass of the offices closing at 5·30 P.M., it will be seen from the table produced at page 231 that the cost per unit is but 2·49d.

In the "shipping" offices—offices usually located on the platforms of goods sheds, for the purpose of dealing with invoices, &c., and, as a consequence, necessitating very late and sometimes all-night attendance—the advantage of the electric light is still greater, for in these offices an abundance of light is a necessity.

Arc lighting is destined to be of great advantage in railway working. In the loading and unloading of stock, and in the marshalling of trains, it is invaluable. To be able to unload a train, load its contents into trolleys ready for delivery, and *vice versâ*, to transfer from the collecting vans to the railway trucks, marshal and despatch them with speed, is *economy in capital, men, material and time*—results which are not confined to the depôt at which it originates, but which to a great extent influence the traffic of the entire system.

The advantage of the arc light in goods warehouses, yards, sidings and marshalling grounds, presides in the simple fact that it is a larger and truer light than can conveniently be obtained from gas. By this larger light work is handled with greater accuracy and greater despatch. To be able to load a given number of trains in, say, three-fourths the time formerly occupied, is tantamount to increasing the capacity of the premises, warehouses, yard, &c., 25 per cent.; or, supposing a depôt required enlarging in order to meet an increased traffic, the introduction of the electric light should meet the contingency, and avoid purchase of land, erection of additional buildings, &c. In like manner, with new buildings, economy in time means economy in capital outlay: the buildings need not be so large, because the traffic can be disposed of in less time.

Probably to no part of a railway system is the electric light of greater utility, or more valued, than in yard shunting, or the marshalling ground. With a good light at the junction points the marshalling process may be carried on with almost the same despatch and certainty as during broad daylight. Risk to life and limb is largely reduced, whilst it has been found to be in no small degree useful in preventing that petty pilfering to which the absence of light is often an inducement. The author believes it to be a fact that the first arc lighting plant installed for yard working—viz. that at Nine Elms, was laid down mainly with a view to suppress certain depredations on valuable goods which the absence of light appeared to foster, and upon which its introduction has had so marked an effect as to help largely to defray the cost of the installation.

The electric lighting plant now in use on the Midland is approximately as represented in the table on the next page.

It may be gathered from the fact that these plants have thus, one after the other, been established, the advantages attending the employment of the electric light are of a marked character. With all direct-current lighting, provision

Engine B.H.P.		Locality.	Arc Lights.		Incandescent Lights.
Steam.	Gas.		2000-C.P.	1250-C.P.	8-C.P.
200	90	Bradford .. ..- ..	..	190	1100
190	..	Leeds (Hunslet Goods Depôt)	..	150	523
190	..	Sheffield (Wicker Goods Depôt)	114	..	300
190	..	Liverpool (Sandon Dock) ..	..	126	100
190	25	Birmingham (Central Goods)	73	..	435
300	..	,, (Lawley Street) ..	..	200	400
320	..	Derby	..	..	4500
..	250	Leicester (Passenger and Goods)	160	..	400
300	..	St. Pancras ,, ,,	242	..	..
..	..	,, Hotel	..	..	1330
200	..	Nottingham	55	75	300
..	40	Wellingboro' .. :	35	..	100
2080	405		679	741	9488
2485			1420		

of some kind is necessary to meet the fall in potential consequent upon increased output of current; whether this can be best accomplished by batteries or by *compensators* has yet to be shown. The author inclines to a preference for the latter, as being less troublesome, more constant in action, more reliable, and—which is a very important question — occupying much less space. Compensators, or as they are there termed "Boosters," are largely used in the United States, and there is little doubt they will shortly find an extended application in England. The author has reason to believe that high-tension direct-current working will prove the system which will commend itself as that most suitable to railway requirements. So far the demand has been for lighting. Lighting will, undoubtedly, always prove a large factor, but demands for power will probably largely exceed those for light. When we realise the fact that from one source—

one generating plant—we can serve the demands of a district, or section of line, for lighting as well as those purposes for which independent steam power or hydraulic plants have now to be provided, we can entertain little doubt of the economy which must attend such a concentration of staff and power. With all such systems *compensators* will prove a necessity.

At the Midland Company's Bradford station the wiring has been carried out on the simple parallel system. At Derby the three-wire system is in use. At each depôt compensators from the design of Mr. J. Sayers are in use. At Bradford the extent of the compensation is limited to 11 volts, with a maximum current of 320 ampères. At Derby a network has to be served and the compensation is more involved, owing to the loss in the middle wire when either of the two sets of three separate groups of feeders is unevenly balanced. Dealing with one set, the arrangement is as follows :—The outside wires of the grouped feeders are connected through one of these compensators, which in point of fact is a series machine through which the entire current for that circuit passes in order that its potential may be raised, and which is driven by motor at a constant speed. This compensates for the loss in each outside wire.

The loss in the middle wire is double-acting, raising the volts on the light load side and depressing them on the other side. To meet this, Mr. Sayers carries the middle wire round the magnets of each of the two compensators, but in opposite directions one to the other. Any current passing in this wire therefore increases the volts in one and decreases the volts in the other, equally ; and by suitably proportioning this winding it can be made to practically neutralise the loss in the third wire.

If both sides are fully loaded and there is no loss, or no current passing in the middle wire, each series machine has to raise the voltage, on a current of 350 ampères in each, some 7 volts ; but, as the third wire is one-half the section of the outer wires, it follows that, if one side should be fully loaded and the other not loaded at all, the compensators have to give $7 + 14 = 21$ volts.

to Railway Working. 219

Such an extreme case has not occurred in practice; still, serious differences do occur—more so, perhaps, in such a supply as that under reference than in a town service—for the reason that, however equally you may plot out your scheme, it is quite impossible to foresee all those contingencies which may call for extra service on the part of a special branch of a railway staff.

A further advantage to be derived from the use of these compensators is to be found in the regular range of speed at which the engines may be worked. The volts of the main dynamos merely require to be kept at the normal pressure. There is no need for running the speed up so as to increase voltage to meet the drop consequent upon increase of current; thus the employment of these machines effects three important services: (1) they maintain an equal pressure; (2) they simplify the working in the engine room; (3) and consequently economise labour, there being no longer that need for constant presence at the steam valve to meet every variation in voltage.

For goods yard and siding working the selection of the positions for the lamp pillars, and the height the lamp should stand from the ground, are matters of importance. The former is only to be dealt with in a satisfactory manner by learning from those who have the handling of the traffic, and by personal observation, where the light is most needed.

Each group of *points* requires special consideration. Shunters should be able to see when waggons have passed over each "point," so that they may readily "knock off" the next wagon or group of wagons as the required "points" are clear of the trucks previously despatched over them.

As an instance of meeting a difficulty where roads are numerous and a good light desirable :—At the mouth of the Leeds Midland goods yard, where there are four roads, a light girder bridge (Fig. 123) has been thrown across, from which three lights are suspended between the roads, and the lighting is in consequence very efficient. The point is an important one, as from each line several fans of roads radiate, and it is beneath this bridge that the main part of the shunting is

Application of Electricity to Railway Working. 221

done. Without the girder bridge it would have been impossible to adequately light the intervening spaces between the roads.

The height at which the light is required for the illumination of a railway yard, or for the lighting of points, will, of course, determine the current required, whether 10 ampères or more. The author believes the 10-ampère lamp will prove most serviceable for this work. The dispersion of light is not required to be general, but local, applicable chiefly to the *points* at which the traffic has to be diverted.

In station buildings, where structural arrangements do not intervene, the 10-ampère lamp may be used with advantage. In goods yards and open spaces, theoretically they should illuminate an equal extent of area. In point of fact, however, the illumination does not bear the same relation when used out of doors as within enclosed spaces. On the Midland it has been found necessary to employ only one class of lamp pillar for outside work. The height adopted is 20 feet—i.e. 20 feet between the arc and the level of the rails. The distance these pillars are apart varies for the 6·8-ampère lights from 90 to 100, and for the 10-ampère lights from 100 to 120 feet.

The distance between the lamps in the goods sheds must, in a great measure, be controlled by structural arrangements, cranes, &c. In open sheds, where these difficulties do not exist, a distance of 45 feet between 6·8-ampère lamps, and a distance of 60 feet between 10-ampère lamps, will usually be found convenient. Every case will, however, call for special consideration. To lay down a rule which shall be universal and at the same time useful is practically impossible.

As an instance in point, the lights employed on the St. Pancras passenger station are 10-ampère. At the head of the station those lamps are 60 feet apart, the space between them increasing towards the end of the platforms, where the passenger traffic is not so heavy, the distance between the two most distant lamps being 90 feet. The height of these lamps from the platform is 14 feet.

On the Leicester passenger station main platforms, where

10-ampère lights also are used, the lights are 90 feet apart, and at an elevation of 15 feet from the platform level. At Leicester the glass roof is lower than the St. Pancras roof, and the walls, being of glazed material, have a much greater reflecting power than is the case at St. Pancras. The side platforms of the new Leicester station are covered by the usual low glass awnings, and the lamps have to be arranged to meet the structure. Here they are 13 feet above the platform, and 75 feet apart. The position in which these lamps are obliged to be placed results in a curtailment in the distance between the lamps. The low awnings are more expensive to light than is the main building.

The height at which arc lights should be arranged in order to secure the necessary illumination, depends upon the positions at which the light is most required and the area. In railway goods yards and marshalling grounds it is chiefly required at the *points*. It is there the work has to be done, and although a certain amount of light may be needed elsewhere it is but small. Cart-ways and approaches, as also those sidings at which goods are loaded into or from wagons, naturally require a good light. Whether this can be best met by large lights placed at a considerable elevation, or by moderate lights at a low elevation—say 20 feet—calls for careful consideration.

Elevated lights to be of any real service must be lights of large power, as those employed at Nine Elms, or those somewhat recently erected at the Great Eastern Company's chief London goods depôt, Liverpool Street. Such elevated lights will of course call for greater expenditure in pillars than those for a 20-feet elevation. Where a powerful lamp will sufficiently light an area which would otherwise call for two or perhaps three smaller ones, there will possibly be a saving in labour, but not much, for the trimmer in dealing with the former has in all probability to ascend the pillar to a certain position, then to lower the lamp, and thereafter to pull it up again and himself descend to *terra firma*. In dealing with a 20-feet pillar he has but to ascend his ladder to reach the lamp, and again descend.

The author inclines to the opinion that the lower elevation will be that which will be found most useful for railway purposes.

In considering this question it may be of interest to reproduce the following data worked out by Mr. L. W. Preece, and which appeared in the columns of the *Electrician* some time since.

"The accompanying diagram is a series of curves showing the illuminating power of a 1000-candle-power arc lamp at various heights. It is based upon Mr. A. P. Trotter's paper

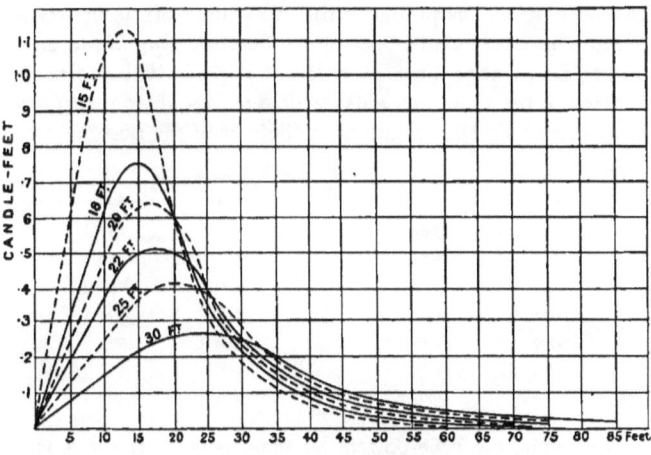

FIG. 124.

on "The Distribution and Measurement of Illumination," read before the Institution of Civil Engineers in May 1892, where it is shown that the illumination of an arc lamp varies as the fourth power of the cosine of the angle of incidence ($\cos^4 \theta$) for angles greater than 45°, and also that the illumination at about this angle is at the maximum. The curves are arrived at from this information, being obtained as follows:—Assuming the light of the lamp to be 1000 candle-power, the illumination underneath the lamp, ignoring the lower carbon, would be 1000 candle-feet divided by the square of the height.

This may be taken as the value of illumination with θ at 0°. The curve, however, only starts at 45°, the value $\cos^4 \theta$ being 0·25. The actual maximum is therefore 0·25, the supposed value under the lamp. Starting from this point, the curves are the various values of $\cos^4 \theta$ divided into the supposed value of illumination under lamp. The ordinate is marked out in feet instead of the angles of incidence, being simpler for practical work. The scales, both for the feet on the abscissa and candle-feet on the ordinate, are quite arbitrary."

Mr. A. P. Trotter,[*] who has bestowed much time and thought upon the subject, in discussing the advantage of various heights, has produced the following table as the result of many measurements, "not of candle-power, but of illumination of an average lamp at different angles. The angles are expressed by their tangents, which represent the distance measured along the ground from the foot of the post."

Tan.	C.P.	Tan.	C.P.
0·0	216	1·4	890
0·2	348	1·6	830
0·4	520	1·8	770
0·6	785	2·0	690
0·8	970	2·5	600
0·9	1000	3·0	507
1·0	985	4·0	415
1·2	940	5·0	308

From these candle-powers he represents by the following diagram the illumination for various heights, due to such a lamp whose maximum at a little less than 45° is 1000 candle-power.

The dots on the vertical line show lamps 20, 30, 40 and 50 feet high. The illuminations are plotted in candle-feet vertically, and distances are plotted horizontally. Mr. Trotter then proceeds, "With a lamp 20 feet from the ground, there is a brilliant maximum at an angle whose tangent is about 0·6. The illumination is very small at an angle whose tangent is 4. If you increase the height of the lamp to 1·5—that is

[*] Institution of Electrical Engineers, April 4, 1895.

30 feet in this case—you get less than half the maximum illumination; but the minimum is considerably better. With a lamp at 40 feet high, the maximum illumination is rather small compared to that of a 20-feet lamp, but it is quite enough for a goods yard; and the minimum is much better. Finally, if you go to 50 feet, beyond which rather serious structural difficulties present themselves, the maximum illumination is small, it is true, but it is spread over a wide area. There is very little variation, and a very good minimum. The illumination due to the next lamp will, of course, double

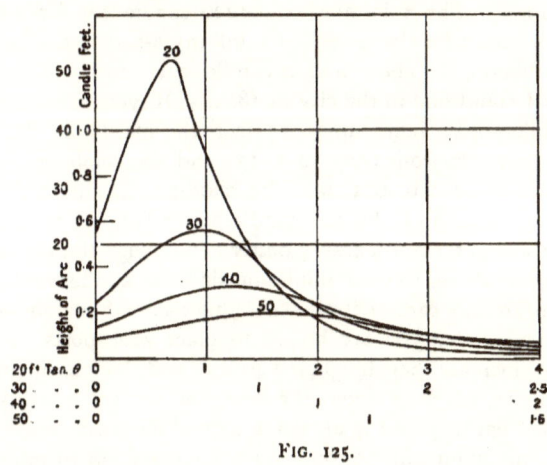

FIG. 125.

this minimum. For ordinary street purposes 50-feet posts are out of the question—there is not enough area for them to command—but they are very useful for goods yards and other wide spaces.

"I have taken the candle-power of the 6·8-ampere lamp with a maximum of 900 candle-power at 45°. You may get more, perhaps, with a laboratory measurement, but 900 is as much as you can get on an average in practice with a globe or lantern. I have taken the 10-ampere lamp at 1,500 candle-power maximum. With these I have calculated the illu-

Q

mination which Mr. Langdon is probably producing. The first two lines of the following table give the maximum and minimum illuminations, using Mr. Langdon's data:—

C P.	Height.	Distance.	Max.	Min.
900	20	90	1·1	0·206
1500	20	120	1·85	0·12
1500	25	120	1·17	0·153
1500	40	120	0·5	0·28
900	40	120	0·3	0·168

"The maximum illumination is about 1·1 for the 900-candle-power lamp, and about 1·85 for the 1500-candle-power lamp. The minimum is 0·206 candle-feet for the former, and 0·12 for the latter. I think my assumed candle-power is liberal, for about half a candle-foot was about the best that I could find in the city in 1892. If you raise the 1500-candle-power lamp up to 25 feet, the maximum illumination falls off from 1·85 to 1·17, and the minimum is increased from 0·12 to 0·153. The maximum is almost the same as in the case of the 900-candle-power lamp at 20 feet high, but you get a considerably better minimum. But if you only want to attend to your minimum lighting, and leave the maximum to take care of itself—as I think should be done—then, assuming that you are bound to place your posts 120 feet apart, increase their height to 40 feet, and the maximum comes to 0·5, which is about the best you can get in fairly good street lighting, and is almost too good for goods yards; and the minimum will be very much increased, up to more than a quarter of a candle-foot. The theoretically best height for a light of uniform candle-power would be 42 feet 6 inches; but for an arc it should be higher, since the maximum candle-power is found at a smaller angle with the vertical. But even if you take the smaller lamp—the 900-candle-power lamp—and put that upon 40-feet posts 120 feet apart, you get 0·3 for the maximum, and you get 0·168 for the minimum; that is to say, the 900-candle-power lamp on 40-feet posts 120 feet apart, will give you better minimum than the 1500-lamp on 20-feet posts 120 feet apart. I think that is worthy of some consideration. These numbers are simply calculated with a slide rule. The third figure has no practical value; no illumination photometer works closer than 2 or 3 per cent."

Mr. Trotter bases his argument upon the assumption that the light required for a goods yard is a generally dispersed light, a light which shall be practically the same throughout. There are points, of course, where a generally fair amount of lighting is necessary, but this is exceptional. The demand is for a good light at the "points," where the marshalling has to be done. This is the chief demand, and Mr. Trotter has shown that with a 6·8 and a 10·0-ampère lamp, the best result is obtained at an elevation of about 20 feet. Mr. Trotter's remarks are reproduced in order that the reader may have the advantage of his arguments in determining the course he may desire to pursue in establishing lights for similar work.

Double carbon lamps are usually necessary for railway working—the hours of lighting during the winter months being long. The size of the carbons most suitable is not quite determined. Various sizes have been employed—from 12 to 18 millimeter for the upper, and 12 to 15 millimeter for the lower carbon. On the whole it is probable the 13 millimeter affords the most light, while the 18 to 13 millimeter effects, possibly, a slight saving of labour during a portion of the year. The employment of the larger size carbon for a 10-ampère current does not, in practice, produce so large an economy in labour as would appear to be the case.

During the summer months the trimming is not so frequent, and the employment of one or more trimmers may, according to the extent of the work to be disposed of, be dispensed with. It is better, however, to take advantage of such periods to employ any spare labour in painting and renovating lamps, lamp pillars, and otherwise preparing for the winter's run, rather than to break up a staff of men well established in their work.

In some of the earlier plants laid down by the Midland, the lamps have been protected by sheet iron hoods A, Fig. 126. The lamp is suspended from a cross piece of timber B by insulated hooks or hangers C, and is further supported at its base by an insulator D.

Later installations have been equipped with four-sided

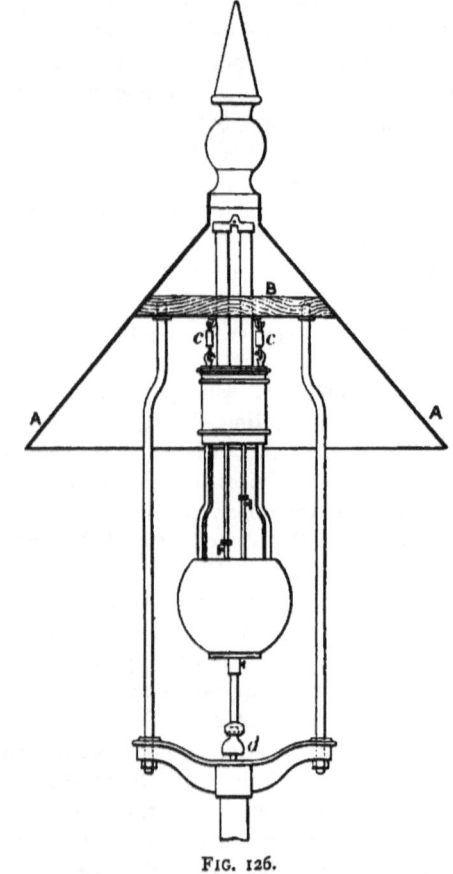

Fig. 126.

lanterns, Fig. 127. The upper portion of the lamp is insulated from the lantern at A by a block of dry wood and insulated hooks, and its base, as in Fig. 123, is supported by an insulator B.

to Railway Working. 229

These lanterns have been glazed—
(a) With plain clear glass;
(b) With what is termed "muffled" glass;
(c) With fluted glass.

Clear glass can be used only to a limited extent. The naked arc is objectionable to men working within its immediate range. "Muffled" glass has been employed with the object of toning down the glare of the arc and breaking up the shadows. It is not, however, in this respect so successful as the fluted glass. With the flutings horizontally arranged the refraction is better.

The lanterns are made so that the chimney may lift off at A. All the lamp pillars are provided at their base with switches, which sever the connection of the lamp wires with the line wires; and, as is desirable, the switch, when in the "Off" position, also places the lamp to earth. The lamp pillar complete is shown in Fig. 128. The switch is enclosed in the base of the pillar.

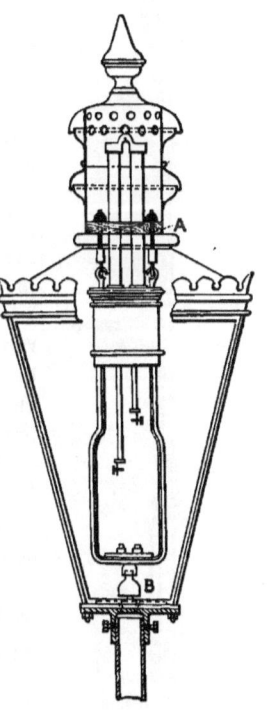

FIG. 127.

Where lamps are worked in series it will, in determining the circuit arrangements, be found desirable to ascertain the period during which the lights are required for use, and to group them accordingly, so that they may be turned off and on in sections. The groups should also be so arranged that, combined, they may equal the capacity of a machine. The circuits applicable to indoor lighting should be so arranged as to admit of their being worked independent of the outdoor or yard circuits, the light being required indoors at an earlier hour than those out of doors.

For incandescent lighting the wires are frequently run in grooved casing. The casing and cover should in all cases be fixed by the outer and not by the centre fillet; the object in rejecting the centre screw so frequently employed, is to exclude the aid it would afford to any *creeping* of the current across the centre fillet, should leakage at any time arise. Under such a condition it is clear the screw in the centre fillet would largely reduce the resistance between the two wires—that it would, in fact, become a sort of "half-way house" towards the passage of the current, and tend to ignite the wood. The wooden casing as at present employed affords a convenient mode of dealing with surface wiring of rooms; but wherever the wires have to be placed out of view, they should be enclosed in metal tubing capable of carrying the maximum current which the wires enclosed within it could carry without fusion. If an artistic rectangular casing of metal which would not readily melt could be produced at a reasonable price, it would no doubt soon supersede the wooden casing, which, although it has so far proved of considerable use in the extension of electrical energy, is scarcely what is desired. Its great recommendation is its cheapness, and the ready manner in which it lends itself to buildings already complete or in occupation. Where wood moulding can be made to harmonise with the decoration of a room, the ordinary metal tubing would, in most cases, be unpleasant to the eye.

FIG. 128.

To lay down an electric lighting plant is doubtless more

to Railway Working. 231

costly than to establish means for illumination by gas; but when we take into consideration the relative cost of working the light derived from each source—light for light, or candle-power for candle-power—the advantage is so largely in favour of the electric light as to more than fully compensate for any larger first outlay which may be incurred.

The following tabulated statement affords data with respect to the output and cost of working of several of the electric lighting stations on the Midland Railway.

MIDLAND RAILWAY—ELECTRIC LIGHTING.

Details of cost, half year ending December 31, 1895, *for maintenance and renewal —exclusive of charges for land, taxes, or aepreciation of buildings—at the undermentioned stations.*

INSTALLATION.	Total Units.	Total Cost.	Cost.						
			Per Unit, including all Charges.	Per Arc Lamp Hour.	Per Unit for Incandescent Lighting.	Per Unit for Labour.	Per Unit for Material.	Per Unit for Repairs.	Per Unit for Coal.
		£ s. d.	d.	d.	d.	d.	d.	d.	d.
Somerstown	205,204	2873 9 8	3·36	1·68*	..	1·13	1·11	0·39	0·72
Leicester	126,418	1738 0 5	3·1	1·55*	3·1	1·49	1·17	0·15	0·41
Derby	81,295	842 13 2	2·49	..	2·49	1·36	0·54	0·13	0·47
Birmingham (Central)	87,017	831 11 8	2·29	1·145*	2·29	1·13	0·76	0·25	0·3
Birmingham (Lawley Street)	123,215	1269 10 1	2·47	0·92	2·47	1·16	0·53	0·44	0·33
Sheffield	131,783	1618 4 11	2·9	1·45*	2·9	1·14	0·88	0·4	0·51
Leeds (Hunslet) ..	81,978	1225 10 6	3·5	1·31	3·5	1·8	0·68	0·61	0·5
Bradford	160,365	1817 18 9	2·7	1·01	2·7	0·92	0·57	0·7	0·34

* These are 10-ampère lamps, 2000 C.P. All others are 6·8-ampère, 1200 C.P.

These charges are in no case comparable with those of city supply stations. The arc lighting load, varying throughout the night according to traffic demands, is not so constant as that of a town service; while that for incandescent lighting lacks the continued evening service associated with shop or residential lighting. At goods stations and for office lighting the latter rises to its maximum in the depth of winter for a period not exceeding an hour and a half daily, viz., 4 to 5.30 P.M.; while in the summer the demand is still less.

It is also to be borne in mind that these results do not cover the outlay for land, taxes, or depreciation of buildings. Nor is anything added for depreciation, for the reason that repairs and replacements are dealt with as they arise, and are thus, year by year, included in the repairs column.

In criticising the foregoing table it is necessary to bear in mind the fact that the cost per unit carries with it the cost of carbons and carboning the arc lamps, together with cleaning, repairing, &c.; and that the cost for incandescent lighting also, as a rule, covers the renewal of lamps. Further, that the appropriation of charges to incandescent and arc lighting where both are in operation, is unavoidably somewhat arbitrary. There are many items of cost which apply to both, such as labour in attendance upon the machinery, cleaning, &c. These have to be divided and appropriated as is considered fair and reasonable. The question will arise, Is one favoured at the cost of the other? The result must speak for itself. Still, whatever we take from one must be debited to the other. If we increase the cost of one, it will cheapen that of the other.

Some of the Midland Company's recently constructed goods depôts afford an opportunity for comparing the cost of the electric light with that of gas. Gas has in some instances been adopted as an alternative form of light in case of failure with the electric light. Consequently, the gas lighting is what may be regarded as modern gas lighting, affording an amount of light which, though not nearly so great as that afforded by the electric light, is greater than that which attained, and still attains, under the old system of gas lighting.

Fig. 129, which is a plan of one of the warehouses, shows the positions of these gas lights, together with those of the electric lights. The former are on the regenerative principle, and arc each nominally 150 candle-power; the latter are arc lights, nominally of 2000 candle-power.

It is officially stated that 5 cubic feet of gas, burnt under the principle referred to, will produce 25-candle-power light for one hour. There are 81 150-candle-power gas lights.

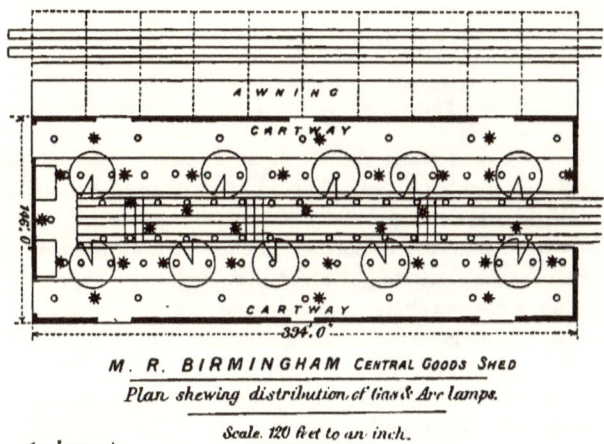

M. R. BIRMINGHAM Central Goods Shed
Plan shewing distribution of Gas & Arc lamps.

Arc lamps ✶
Gas lamps ○

Scale. 120 feet to an inch.

FIG. 129.

The price of gas is 2s. 3d. per 1000 cubic feet, less 5 per cent. On this basis the cost of gas lighting would be 62d. per hour.

There are 30 arc lights, the cost of which, as per the foregoing table, averaged throughout 1·30d. per light per hour. The total cost per hour would therefore be 39d. as against 62d. for gas; the relative nominal candle-power being—

Gas	12,300
Electric light	56,000

At another depôt there are 53 arcs of 1200 candle-power, and as an alternative, if required, 130 gas lights of 150 candle-power each. The arc lights here cost ·92d. per lamp hour, but it will be better to adhere to the average cost, viz. 1·3d.

Gas is 2s. 3d. per 1000 cubic feet, less 5 per cent. The result works out per hour— .

	d.
53 arcs, affording 63,600 C.P.	68·9
130 gas, affording 19,500 C.P.	100·00

At a third depôt the gas used is manufactured at the company's own gas works, and is priced at 2s. per 1000 cubic feet. The result is as follows :—

	d.
29 arc lights, affording 58,000 C.P., cost per hour,	37·7
86 gas lights, affording 12,900 C.P., cost per hour,	61

If we extend these results so as to cover a period of a year—say 3000 hours—we have a saving of nearly 1000*l*. per annum on the three ground floors of the goods warehouses referred to ; and it is important to observe that, whereas the charges for the electric light cover all running charges—repairs, cleaning, &c.—the charge for gas is for gas *pure and simple :* there is no provision for cleaning or for repairs.

The Midland has also laid down for electric lighting a gas plant at Leicester. This plant consists of four gas engines, required to deliver under Dowson gas 40 horse-power, and under coal-gas 50 horse-power, on the pulley of the dynamo, for the purpose of driving three 50-light 2000-candle-power Brush series high-tension dynamos, and one Brush low-tension dynamo employed chiefly for arc lighting ; and two 14-horse-power nominal gas engines, required to transmit 25 horse-power to the pulleys of two Siemens incandescent machines.

The service pipes are in duplicate : that is, one is in connection with the town gas, the other in connection with a Dowson gas plant. The engines are provided with valves which admit of the gas supply being changed over at will from coal gas to " Dowson," or *vice versâ*.

In construction, an ordinary gas plant saves the cost of boilers and chimney shaft, and occupies less space, thereby effecting an economy in cost of building and cost of land. No stoking is required when working off coal gas ; and, as a rule, the attendance upon engines and dynamos is shared by the electrical attendants—engine drivers and stokers are non-existent.

In that it is at any moment ready for use, it possesses a certain advantage over steam plants, which are not always *in steam ;* it is less costly in labour, but it is not so compact as many steam plants, does not admit of direct driving, is more noisy, and requires careful attention to wearing parts. It is unsuitable for driving alternating-current machines in parallel.

The experience so far gained establishes beyond doubt advantages largely in favour of the electric light as an illuminant for dealing with railway work wherever a large body of light is required.

At the same time, daily experience teaches us to look for further developments in the machinery employed. One improvement should be the abolition of belt-driving. All driving should be direct from the shaft of the engine. The result will be less danger of interruption, less space, greater regularity.

A further advance is to be looked for in the unification of the class of generator. At the present moment it is necessary, where series-arc and incandescent lighting has to be provided, to lay down independent power and independent machines for each, or to drive the arc lights from alternating or transformed alternating currents.

What is needed is a generator the energy derived from which may be applied alike to incandescent lighting, arc lighting, or motive power.

Few, probably, will doubt that electrical energy is destined —and that at no distant date—to become a useful and economical agent at all important railway centres, not merely for lighting, but for other purposes at present met by horse labour, hydraulic power, &c. In determining the ultimate design of generator and mode of working, this probable demand should be present to our minds. That the demand for electrical energy will increase we may be sure, and the provision first laid down should in all cases, both in regard to buildings and machinery, be capable of extension.

Another point to be achieved is the concentration of the generating power; the generation from one spot of that necessary to meet the demands of a section of line, say five miles. At busy centres such an installation will prove of the utmost value. The Great Northern Railway has been the first to take this step in relation to lighting. At their Holloway Station, a fine building, capable of large extension, has been erected, and plant laid down for driving both arc and incandescent lights for such a section of line as has been referred to above, including the King's Cross Station, offices, hotel, goods depôt, &c. The Midland also are taking a further step, in the erection of a central generating station near Kentish Town, for the purpose of meeting not only lighting but power demands between Welsh Harp and St. Pancras. This central station will generate current at 2000 volts, which will be transformed down to 220 volts at such points on the way as may be necessary by rotary transformers. Direct current has been selected specially with a view to motor work and arc lighting. It is believed that greater economy will, having regard to the irregular demand to which the machinery may be required to submit when working pumps, lifts, capstans, &c., be achieved with this system than with alternating machinery and apparatus. The unit will be 200 kilowatt. Lancashire boilers are to be used.

In employing high-tension currents on railways, it is desirable the cables employed for conveying these currents should be laid in such positions or in such a manner as will render it practically impossible for the current to be communicated to a platelayer or other person who may possibly come into contact with them, either directly or by means of any implement which they may be handling. Concentric cable should be employed, and in carrying it along the line it should be laid at such a depth as will place it out of the general run of the platelayer's pick. It is doubtful if so much protection will be afforded it if laid in pipes or culverts, or if covered by flag-stones or bricks, as would be secured by laying the cable beneath creosoted sleepers. The sleepers should not lie immediately upon the cable. The cable should find its bed in some soft ballast, and be covered by similar material some three to four inches before the sleepers are laid over it. As it is undesirable the creosote oil should come into contact with the insulation of the cable, the sleepers employed should be void of any free or surplus creosote. Sound second-hand sleepers would serve the purpose very well. Where such cables are required to be carried along retaining walls, they should be encased in strong wooden boxing, fixed at an elevation above that at which, should a vehicle leave the line of rails, it might come into contact with the wall.

Cables which are not laid in pipes should be "armoured"; and in all cases where they are carried under roadways, or lines of railways, they should, whether armoured or not, be protected by strong cast-iron piping. This is especially desirable where heavy loads are liable to pass over them. The armouring and the lead sheathing are, under such circumstances, liable to become depressed, or squeezed out of position, and thereby to cause decentralisation of the conducting wires—resulting, sooner or later, in failure.

The question of the best form of cable for use with high-tension currents is becoming one of great importance. In some instances india-rubber has failed to afford that satisfaction which was so generally anticipated. Composition insulation encased in lead, is now receiving considerable attention. Its value as an insulator is greatly, if not entirely, dependent upon the power of the lead sheathing to prevent the encroachment of moisture or the disintegration of the composition. How much, therefore, depends upon the prevention of any destruction of the lead chemically, electrolitically, or mechanically will be clear. These are all enemies to the life of the cable, and points which should be present to the mind of those entrusted with the important duty of placing the cables in position.

CHAPTER XI.

TRAIN LIGHTING.

ALTHOUGH at the present time, by far the greater number of railway trains are lighted by means of compressed gas, a very general opinion exists that ultimately, and that at no very distant period, electricity will become the illuminant; and there are not wanting many who at the present moment would greatly welcome its adoption.

Certain railway companies in England, notably the London Brighton and South Coast, the Midland and the Great Northern, and others at an earlier date, have not shown themselves indifferent to the evident desire of the travelling public for electrically lighted coaches. Experiments to a greater or less extent have been undertaken; but the difficulties attending the keeping together of a partially fitted stock, combined with the interruptions and irregularities produced by the not unfrequent interposition of non-electrically fitted vehicles, compared with the ease attending the insertion of gas-lighted carriages, has, with one or two exceptions, created in the official mind a preference for the latter.

On the Brighton line there are, however, at this moment, over forty trains travelling between London and Brighton and other points electrically lighted. On the Great Northern and on the Dublin and Belfast line, trains electrically lighted are also in use. The whole of these trains are, however, trains which run *en bloc*, i.e., are not broken up on the journey.

Some of the first experiments in train lighting by electricity were carried out by the Lancashire and Yorkshire Railway, by means of a small auxiliary engine placed on the locomotive. A somewhat similar attempt was made by the

Great Eastern company. In each case the experiments were not attended with success.

In 1881 a Pullman car was fitted with accumulators and employed in regular traffic by the Brighton company between Victoria and Brighton. The cells were charged by a stationary engine and dynamo laid down at Victoria Station for the purpose. From this the company has continued its efforts, until the number of trains electrically lighted now number, as stated, no less than forty.

To Mr. Houghton, the Brighton company's electrical engineer, and to Mr. Stroudley, must be ascribed the credit of being the first to apply in a practical manner electricity to this purpose.

Where trains run *en bloc* the lights can be served from one source. A dynamo driven by belting, or more direct means, from the axle of the vehicle, together with certain accumulator cells, is placed in the guard's van. Automatic apparatus is provided, by which the direction of the current is controlled and the dynamo connected to the batteries when it has attained the necessary speed, and has, consequently, the requisite electro-motive force. The dynamo, the cells and the lamps are all arranged in parallel circuit—two main cables being led from the dynamo throughout the train. The system of lighting is practically that usually adopted in stationary or domestic installations, plus the necessary automatic apparatus for controlling the current evolved by the dynamo under the varying speed and direction of the train.

To meet the entire requirements of railway traffic, it is necessary that each vehicle should be self-contained, i.e., equipped with generating gear and batteries; or, in other words, that the following conditions should be complied with.

(1) That each vehicle shall carry its own lighting power, that is to say, wherever it may be, whether coupled up to other vehicles or standing alone, it shall be capable of being lighted up as required.

(2) That each vehicle shall be capable of being coupled up, uncoupled, and turned about as may be needed, without the

possibility of the connections being wrongly made, and in such a manner as may be readily understood by those whose duty it may be to make up the trains.

(3) That the coupling shall readily draw apart when the train is parted, or when provision is required for slipping coaches *en route*.

(4) That the coupling when drawn apart shall, if the light should be off, automatically turn it on, so that no inconvenience may arise from severing the connection.

(5) That each compartment shall carry not less than two lights, each independent of the other, so that in case of either failing one may be available.

(6) That the lighting throughout the train shall be capable of being turned on or off as may be required from any one of the guards' vans.

The requirements of a train dynamo driven from the wheels of a vehicle vary according to the system of illumination which is to be adopted. If it is intended to work direct from the generator, that is to say, without the intervention of accumulators, there appears to be no reason against the employment of a compound wound machine, adapting itself to the number of lights in use up to its maximum power.

In all cases where accumulators are employed, whether in a body, as in the London Brighton and South Coast Railway Company's system, or distributed throughout the vehicles of the train, as in the experiments conducted on the Midland main line trains, it is necessary that the following conditions should be fulfilled.

The dynamo should automatically come into circuit when the necessary speed to enable the electro-motive force of the machine to overcome the back electro-motive force of the batteries has been attained. Thus, if the number of cells in series is twenty, the electro-motive force of the machine must be raised to 50 volts before coming into circuit with the batteries.

The electro-motive force should remain practically constant, whatever may be the increased speed of the train. It is found in practice that manufacturers are able to provide this in "thirds"; that is to say, the full electro-motive force is attained at one-third of the maximum speed fixed for the running. In other words, if the pressure is fixed at 50 volts, and the limit of speed at 75 miles per hour, the machine should develop its maximum, viz. 50 volts, when the vehicle is running at 25 miles per hour, and continue to give the same electro-motive force up to a speed of 75 miles per hour.

The current must be in one direction, whichever way the vehicle is running. This necessitates an automatic reversing switch, governed by the motion of the dynamo or of the wheels of the carriage.

Practically speaking, the electro-motive force of an ordinary separately excited dynamo varies directly as the speed at which the armature revolves, so that if the number of revolutions be increased, say from 500 to 1000 per minute, the electro-motive force will be doubled. But the electromotive force varies also directly as the strength of the magnetic field.

If, therefore, the strength of field of a separately excited dynamo be reduced in the same proportion as the speed of the armature is increased, then the electro-motive force of the current will remain practically constant. To effect this, Messrs. Holmes, of Newcastle, arrange two dynamos, A and B, Fig. 130, having their armatures upon the same shaft so as to revolve together, but influenced by separate magnetic fields. A is the main generating machine, the strength of the magnetic field of which is regulated by B. The magnets of both dynamos are separately excited from a set of accumulators C, those of A being provided with two distinct exciting circuits. One of these is an ordinary high-resistance shunt circuit, and the other, which is of less resistance, is coupled up to the source of current so as to have the small regulating armature in series with it. The regulating armature is so connected that its electro-motive force opposes that of the external source of supply. The high-resistance shunt circuit is of such proportions that, at the maximum speed at which it is intended to run the machines, it will give, without the aid of the regulating circuit, a magnetic field of an intensity proper for the required

R

electro-motive force in the generating armature. When thus driven, the second exciting circuit ought to have no current passing through it, a result which is brought about by the electro-motive force of the regulating armature being, at its highest speed, equal and opposite to that of the accumulators, which provide the current for exciting the magnets. If, however, the speed falls, the electro-motive force of the regulating armature is lessened in the same proportion, and being no longer equal to the external pressure, permits a current to flow through the second or low-resistance exciting circuit of the magnets of A, thus increasing the intensity of the magnetic field to make up for the reduction of speed. The result is that a practically uniform electro-motive force is maintained. When the speed falls to a lower limit, say one-half the highest rate, the electro-motive force of the regulating armature is half that of the accumulators, so that it is necessary to calculate the proportions of the second exciting circuit of the magnets upon the basis of only half the electric pressure of the batteries being available, and the resistance of the regulating armature as forming part of the circuit resistance. At the lowest speed, therefore, half the exciting power will be produced by one coil having the full difference of potential of the accumulators at its terminals, and the remainder by the second coil of lower resistance, having half that pressure at its terminals, less the fall of potential due to the resistance of the regulating armature which forms part of this circuit.

A centrifugal governor on the armature spindle automatically switches on the storage batteries at D for exciting the magnet coils of both the generating and regulating machines, as soon as the dynamo attains the minimum speed at which it is intended that it should come into action; and in like manner also switches off the current when the speed falls below the minimum. Carbon brushes placed radially to the commutators, the lead being automatically altered to suit the direction of rotation, have been found serviceable.

On the spindle carrying the armatures is arranged a reversing switch F, actuated by friction discs which come into contact when the spindle slows down, but run free on its attaining any degree of speed. The duty of the reversing switch is to ensure that the current shall always be in one direction, whichever way the vehicle is running.

H is an automatic cut-out, such as is generally used in electric lighting installations, having its high-resistance coil as a shunt to the dynamo terminals, and the low-resistance coil in the main charging circuit. The duty of this instrument is to disconnect the dynamo from the charging circuit at all times when its electro-motive force falls below that of the accumulators employed throughout the train. The weight of these dynamos with bed-plate is 15 cwt., and the weight of the counter-shafting, pulleys, &c., 13 cwt.

Figs. 131A and 131B, which represent a portion of a guard's van, illustrate the arrangement adopted in carrying out the Midland Railway experiments. Upon the axle A, Fig. 131B, are keyed two driving pulleys B B, 21 inches in diameter, and upon the countershaft C corresponding pulleys D D and E E, 16 inches in diameter, which transmit the power to the 9-inch pulleys F fixed upon the dynamo shaft. The dynamo is shown at G.

Duplicate driving gear should be employed in order to meet breakage or other mishap to the belting of either pulley. Where the construction of the frame of the vehicle will admit longer *centres* than are here shown, it is of course desirable such should be employed.

It is false economy to restrict to too great an extent the size of the accumulator cells, and to trust to the dynamo to compensate for any deficiency due to them. They should be of such capacity as will meet the demands of the lamps quite independent of the dynamo. Accumulator batteries are now being produced which will withstand a much heavier discharge without injury than hitherto. The Plauté type is the best for this work. With main line trains, where each vehicle carries its own batteries, the accumulators may be of a smaller type than where one set is provided to serve the whole train.

In arranging the accumulators in the train it is desirable, in order to meet the concussions due to shunting, &c., that the plates forming the cell shall be longitudinal to the train. If this is not done, it is possible that in time the sudden shocks

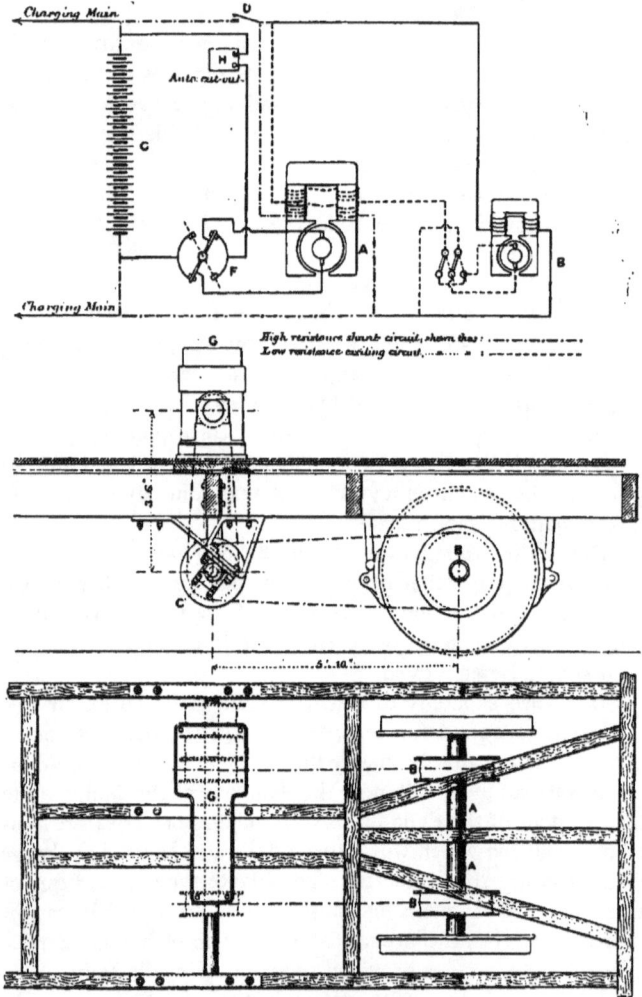

FIGS. 130 AND 131A.

due to shunting, &c., will cause the plates to break away from the shoulder piece to which they are attached. Another necessary provision is that there shall be proper ventilation, not only of the closet, but of the cells, in order to avoid the danger of an explosion from the accumulation of hydrogen.

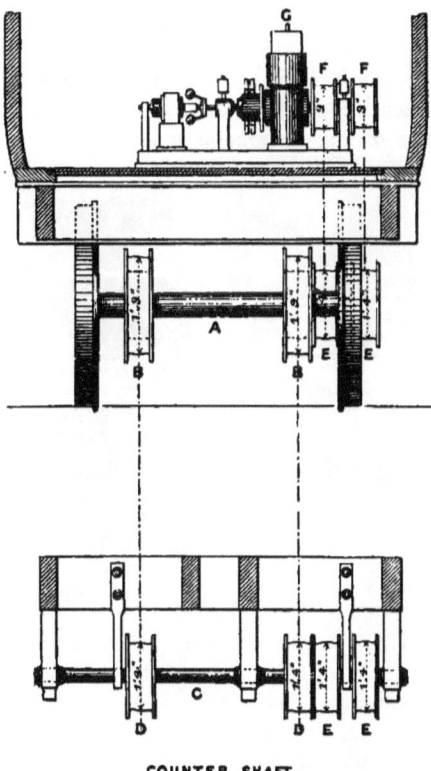

COUNTER SHAFT.

FIG. 131B.

Possibly one reason why electricity has not been more readily applied to the lighting of railway trains is the difficulty of providing an efficient coupling. With trains which are not broken up this is unimportant; but where the marshalling of carriages has to be smartly dealt with at terminal or intermediate stations, it is essential that the coupling should not only provide for a good electrical connection, but also that it should admit of easy handling without the possibility of a mistake.

Fig. 132 is a sectional side elevation, and Fig. 133 a perspective view of one half of a coupling suitable for the purpose. Each portion of the coupling is composed of three forked-shaped bars, Fig. 132; the projections $a\ b$ and $a^1\ b^1$, are split for the purpose of securing elasticity, interlocking the one part with the other, and forming a reliable electrical contact between

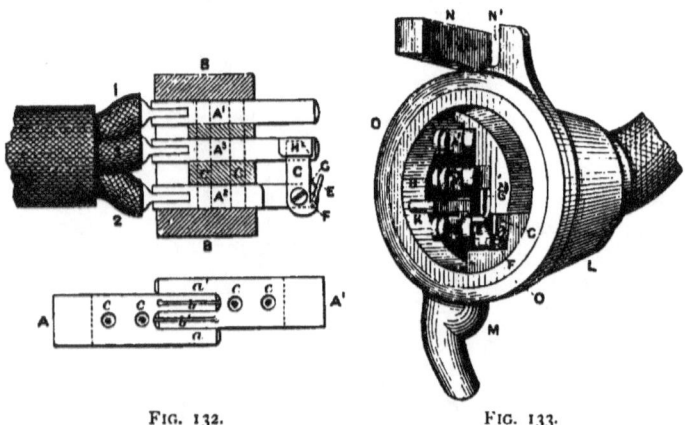

FIG. 132.   FIG. 133.

the opposite halves A and A^1. Three of these bars are coupled together by insulating pins passing through the holes c, and are then inserted in a block of ebonite B. To the several bars $A^1\ A^2\ A^3$, Fig. 132, the main cables are connected. A^1 and A^2 are joined to the cables forming the poles of the batteries, and A^3 to that forming the lamp circuit. A^2 has attached to it a movable connecting bar C which serves to join A^2 and A^3 when required, and thus complete the lighting circuit. C is provided with a trigger piece E pivoted at F, free to be pressed inwards towards C, but otherwise retained in the position represented by means of the spring G. Beyond its fulcrum F, this trigger piece is provided with a shoulder, which limits its movement away from C beyond the position desired. In Fig. 134 is shown a hooked rigid bar K, the object of which is to operate the connecting piece C. K passes over the trigger E, and forces back the connecting bar C until it rests in a recess arranged on the bar A^2. On the two halves of the coupling being drawn apart, K engages with E, and thus draws the bar C over from the recess in connection with A^2, to the shoe H^1 in connection with A^3. In this way, when the two halves of the coupling are pressed together, the cross-connection C is broken and the three bars $A^1 A^2 A^3$ are placed in contact, thus establishing the continuity of the several cables.

The position of K in relation to C, when the coupling is closed, is that represented in Fig. 133. The hook K is not in electrical connection with either of the bars, but is fixed to the block of ebonite in which they are bedded. For slip coaches a special flat bar, not provided with a hook, is necessary, the object being that the contact piece C may be thrust back, but not again drawn over, so as to join A^2 and A^3 when the coupling is pulled apart.

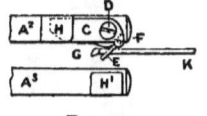

FIG. 134.

Fig. 133 shows the general arrangement, which, externally, is similar to that of the automatic brake coupling. M is the hook at its lower part, and N the straight bar, with a slot N^1 so arranged that the bar N of the one half-coupling shall enter the slot N^1 of the other half-coupling, and the hook M of the one engage with the corresponding part of the other. O is an india-rubber ring fixed in an under-cut groove formed in the flange of the metal case L, serving to exclude the wet. The weight of the two, when hooked together and left suspended between the vehicles which they connect, draws them firmly together at the upper part.

Fig. 135 represents in sectional side elevation the half-coupling joined to the connecting box P, fixed on each end of the vehicle. The three cables g from the vehicle are brought up and attached to terminals h fixed to an insulating disc J.

To a continuation h^1 of these terminals are attached the three leads g^1 comprised within the combined cable Q, and which are at their other ends attached to the bars A^1 A^2 A^3.

At each end of the vehicle it is necessary that a dummy coupling, which, as is implied by the name, is not electrically connected, should be provided. It is composed of a block of wood to which is fixed the hook K, and is intended to provide a position of rest for the coupling when the train is broken up,

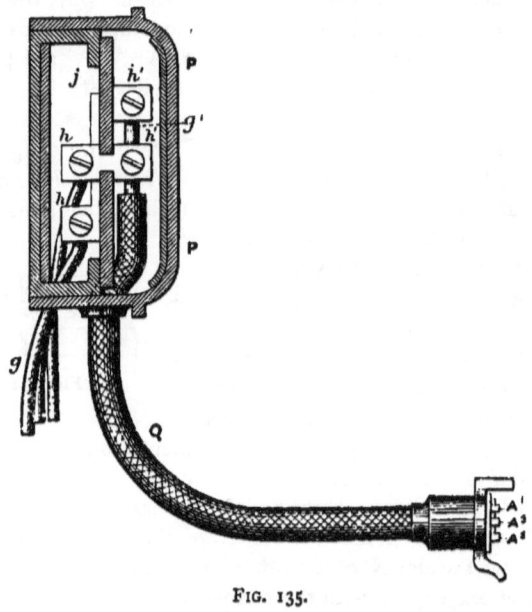

FIG. 135.

and to act as an automatic switch. On the coupling being placed upon it, the connecting piece C, Fig. 134, is thrust back by means of the hook K, and the lights are thereby turned out. If they are required to be retained, it is only necessary to allow the coupling to hang down instead of placing it upon the dummy.

In order to compensate for the increased pressure when the dynamo is in circuit, each vehicle should be provided with an automatic relay which, when the voltage rises above what is required for the lamps, should throw a compensating resistance into the circuit.

With the object of distributing the current equally throughout the several parallels in the train, a small permanent resistance may also be placed in each battery circuit. The output of the machine is scarcely affected by this arrangement, which besides insuring that each set of batteries or lamps receives its

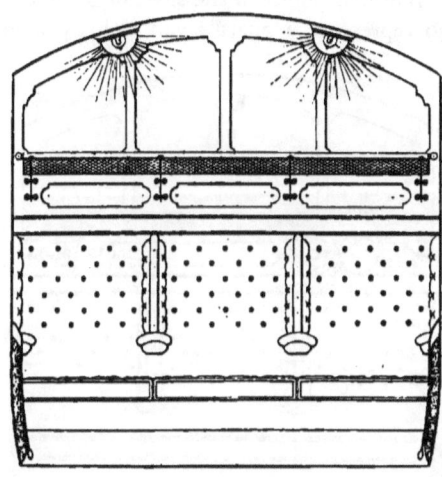

FIG. 136.

proper share, tends to prevent the passage of too great a current from a highly charged set of accumulators to a set in an exhausted condition.

Two lamps should be provided to each compartment. They should be arranged in parallel so as to be independent one of the other. They may be placed in the roof (Fig. 136), halfway between the centre and the doors; or immediately over the two middle seats, i.e. in the direction of the length of the vehicle; or, as in Figs. 137 and 138, between the divisions. In the latter position the occupants have the light at their back, a convenient arrangement for reading; or, where the luggage rack can be dispensed with, they may be arranged at a slightly higher elevation—say in the centre of the panels seen above the position occupied by the lights in the illustration, with reflectors inclined so as to throw the light down upon the occupants of the seats on that side of the compartment, a method which would tend to obviate any glare of the light in the eyes of passengers seated opposite. Each compartment is thus provided with light equivalent to sixteen candle-power, distributed from either two points in the roof or from four points of the sides of the compartment.

Fig. 139 represents a guard's van with dynamo and one

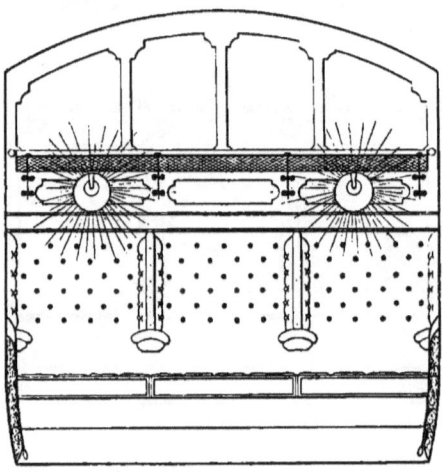

FIG. 137.

coach. From the two poles of the dynamo proceed the main cables P and N, which are continued by means of the couplings between each coach throughout the train. Similarly, a third main L—the lamp main—is carried, from the switch S in the guard's van, throughout the train. The switch S provides a means by which the lighting main L may be connected to the positive main P. Between the charging mains P and N are arranged the accumulators C C", and between the negative pole and the negative charging wire is placed the automatic relay G, the coils of which form the permanent resistance of 0·25 ohm already referred to. The same wire which joins this relay to the batteries makes connection with the lamp circuit, the opposite pole of which is joined to the lamp main. Safety fuses are arranged as shown at F.

On the dynamo being brought into circuit, the current distributes itself by the mains P and N through the accumulators C C″, in proportion to the resistances of the several circuits formed by the batteries. This resistance is that of the cables plus that of the coupling, relay, coil and batteries ; and practically it is the same in each of the parallels. If more coaches are attached, the resistance at the poles of the dynamo is reduced and the current output increased, thereby meeting the additional demands of the train, and this continues until the limit of the power of the machine is reached. The current thus distributed simply passes along the battery mains, through the coils of each automatic relay, and through the three sets of batteries in parallel. The lamps are not affected.

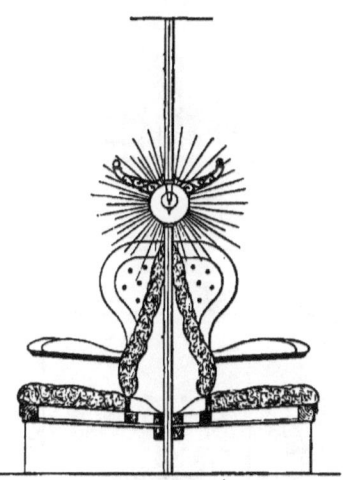

FIG. 138.

If the switch S is turned over so as to connect it with the charging main P, another road is opened, and the current now flows partly round the battery circuit and partly round the lamp circuit of each coach. Starting from the positive pole, one portion passes along P, through the battery lead pp to the batteries, thence by the automatic relay to the negative charging main N ; while the other portion, leaving the main at S, passes by way of the lamps to the resistance controlled by the automatic relay G, through the coil of which it proceeds to the negative pole of the dynamo. The duty of the automatic relay G is to throw into the lamp circuit such resistance as is needed to reduce the electro-motive force evolved by the dynamo to that required by the lamps. This is effected by adjusting an electro-magnet actuated by the current from the dynamo, so that on the armature being attracted, the necessary additional resistance is introduced.

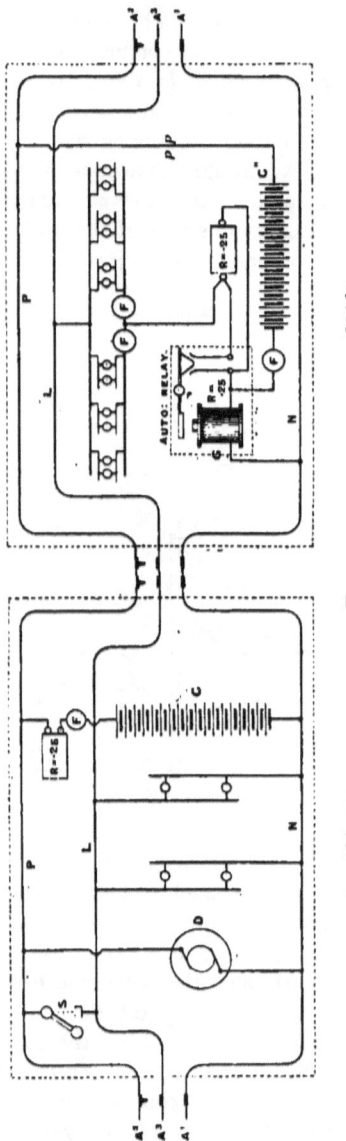

Fig. 139.

When the dynamo is not in circuit, and the light is required, the switch is placed in the same position, and each set of batteries forms its circuit through the charging main P, the switch S, the lamp main L, and the lamps, to the opposite pole of the battery. When a coach is disconnected, the lamp main L and the charging main P are automatically joined at the coupling, and the current from the accumulators circulates throughout the lamp circuit in a similar manner.

It has already been

pointed out that where trains are not required to be broken up *en route*, the lighting can be carried out from one *plant* located in the guard's van or other vehicle. The dynamo need not be of the same description as that referred to. A *Brush* dynamo, and a *Gramme* have been used, but it is necessary that its provision should be such as to ensure a fairly steady potential. This is sometimes attained by employing a machine which readily attains the required pressure, and thereafter maintains it—the magnets having become saturated and incapable of further excitation.

There is no reason why a small auxiliary engine, with dynamo coupled direct, should not be worked by steam from the locomotive boiler, provided such can be afforded a position on the locomotive or tender such as would admit of due observation when running, if required. To effect this, it is needless to say, there must be hearty co-operation on the part of every branch of the locomotive service. The auxiliary engine and dynamo must be within easy reach, and perhaps no better position could be found for it than on either side of the tender, near the foot plate of the engine. Space, weight, and capacity for self-starting are important factors. Experiments which have been undertaken in this direction show that the space required is, practically, 5 feet by 3 feet by 7 feet; the weight of the machine and engine would vary from 12 to 17 cwt. The employment of such an arrangement at once dispenses with those adjuncts needed to meet varying speed and direction of rotation. Moreover, the moment the locomotive is coupled on to the train, the engine may be started and the current generated.

In the early part of this chapter it was pointed out that, in order to meet all the requirements of a train service it was desirable each coach should, in regard to lighting, be self-contained. The most recent attempt in the direction of railway train lighting seeks to make each vehicle not only self-contained but perfectly independent of its neighbours.

Under the system previously described, it was sought to help an infirm vehicle by means of those to which it was connected, so that, if the batteries of one coach became exhausted, the current from those in other coaches should supply the deficiency of the exhausted battery. Messrs. Stone & Co., of Deptford, however, have conceived the idea of placing upon each coach both generator and batteries, with certain necessary adjuncts. They abandon those provisions usually adopted for compensation of the varying rate of speed, and endeavour to meet the difficulty by a *slip of the belt*, which occurs when the train reaches a certain degree of speed.

A small dynamo, together with a pulley to drive it, is mounted on a frame. This frame is affixed to or suspended from the under portion of the coach. The dynamo is driven by a belt from a pulley arranged on the axle of one pair of the wheels of the coach, and the belt is given a degree of tension under the influence of a lever weight in connection with the dynamo frame. Reversing gear is, of course, provided, together with a set of small accumulators, the professed purpose of which is to maintain the light during such time as the train is standing at a station, or until it has attained the requisite speed to generate the necessary E.M.F. to overcome that of the batteries.

As this system is but in its infancy, it would be premature to express an opinion upon its possible success or otherwise. One point in its favour is its individuality. Whatever may go wrong, it will only affect the coach in question.

The Brush Electrical Engineering Company have supplied numerous dynamos for train lighting, and somewhat recently have devoted considerable attention to the question, with the result that they have produced a train-lighting dynamo, illustrated by Fig. 140, which is practically self-contained. The machine is driven from two pulleys keyed on the armature shaft, by means of belting from two similar pulleys arranged upon the axle of the guard's van. When the train starts, the brushes are automatically, by means of a special lever arrangement designed by Mr. Houghton, of the Brighton Company, brought to their proper position, and by the same means, when the train is brought to a stand, thrown out of

to Railway Working. 255

contact with the commutator, thereby leaving the lighting to be accomplished entirely by the accumulator batteries. At the same time that the brushes are brought into position on the commutator, connections are made, so that a current passing round the coils of the field magnet shall afford that polarity required to meet the directron of rotation.

By the aid of the centrifugal governor seen to the left of the illustration, when a given speed has been attained, a circuit with the accumulators is formed through a relay, which latter completes the circuit of the field magnets and so

FIG. 140.

produces their excitation. The charging circuit is then closed by means of an automatic switch, and the charging of the cells commences. Automatically, also, a suitable resistance is inserted in the lamp circuit to compensate for the higher voltage in operation in charging the cells. A small resistance is also arranged for use in the field magnet circuit of the dynamo when the cells only are being charged, and which is short-circuited by the guard when the lamp circuit is closed.

It is understood that several of these machines have been installed on trains in the Australian colonies—the trains running throughout their journey *en bloc*—and are said to have afforded great satisfaction.

The chief drawback to electric lighting for our train services is the source of energy. So far this has been obtained from the wheels of the train. When the train stops the dynamo stops, and the lighting has to be carried on by the batteries. If these are all well charged and in proper order the light will be good, but if the batteries of one coach are, from any cause, run down, not only will the lighting of that coach be bad, but the condition of the batteries will affect the distribution of the current from the dynamo when it comes into circuit. With a partially fitted stock, with certain exigencies in the matter of traffic, and mechanical casualties which are sure to arise with vehicles from time to time, considerable difficulty will attend any such partial application; that is, so far as it applies to trains which have to be broken up *en route*. Naturally, if the entire passenger rolling stock were fitted for lighting electrically, those difficulties would not be so frequent, and, under any circumstances, would be very readily met. But in no case can the electric lighting of trains be carried out so efficiently with the source of energy vested in the wheels of the train, as when derived from an independent source.

This was perceived in the Midland Company's experiments, and to meet it, dynamos coupled direct to auxiliary engines were mounted upon the extreme end of the locomotive tender, and served with steam from the locomotive boiler. The position assigned was not all that could be desired, but it was the only one which could be accorded. A position such as that which has already been indicated would undoubtedly have been attended with less oscillation, but it is a question which has never, so far, been decided, whether the motion of a locomotive would, in course of time, prove a source of trouble to such apparatus.

The warming of railway carriages has, for many years, been carried out in a somewhat crude and cumbersome manner, involving considerable outlay in material and labour, not in all cases to the comfort of the occupants of the vehicles.

Attempts have been made to effect an improvement by passing steam or hot water through pipes carried through the carriages. It is to be feared that the couplings between the coaches, especially when additional vehicles have to be interposed, or the train in any way remarshalled, will give rise to difficulties.

In the United States the tram cars driven by electrical energy are warmed by the same agent.

It would appear worthy of consideration whether, combining lighting and heating, it might not yet be advisable to attach to each train a vehicle equipped for the double duty, having for its primary source of power an oil engine. It is believed that train lighting has already been carried out on a similar basis on the through trains of the Cape railways, if not in America as well.

Such an arrangement would possess this advantage. It would be independent of, and would consequently not detract from that other source of power required for the haulage of the train; it might be maintained in constant operation whether the train were standing at a station or in motion, and it would dispense with the greater portion of those automatic arrangements necessary in electrically lighted railway trains to meet the varying speed and direction of the train.

But it would involve the provision of special vehicles, and probably the services of a man or a youth in attendance upon each; against which would have to be placed the cost in labour and material incurred under the existing mode of lighting and warming.

CHAPTER XII.

INTERCOMMUNICATION IN TRAINS IN MOTION.

ELECTRICAL intercommunication in trains in motion has made practically no progress. It is in use on the South Eastern, the London Brighton and South Coast and the Hull and Barnsley railways. Otherwise the "cord" communicator serves to comply with the requirements of the Board of Trade. The electrical communication in use is practically that indicated in the author's previous work, and consists of a battery in each guard's van, with the necessary apparatus for claiming attention when brought into action. The sets of batteries being in parallel, and the conducting wires at either end of the communication being disconnected, we have an open circuit with the batteries in contention, and consequently so long as the conditions between the batteries remain equal, no current will flow until this equilibrium is disturbed. This is effected by short-circuiting the line, or conducting wires, by turning a switch in the respective compartments of the vehicles composing the train; or by employing a coupling capable of doing so on the severance of the train. The bells throughout the train are thus set ringing. Each guard's van, and if necessary the locomotive engine, is provided with a "plunger" or contact maker, by means of which the officials in charge of the train are able to interchange signals.

In considering this subject we have to bear in mind the object for which the "communication" is required. It was primarily called for with the view of affording a passenger that assistance needed in case of outrage or distress such as would call for the stoppage of the train; also in order that the guard might be able to attract the attention of the driver. To what extent this has been accomplished is somewhat doubtful. At the time the demand for such a means of communication arose, the use of the automatic brake was not compulsory, and was consequently but partially applied. All passenger trains are now fitted with it, and if necessary, as an extreme measure, failing to secure the attention of the driver of a train, a guard may do so by applying the brake. Such a course is not desirable, but it is a means available in emergency. It, however, makes no provision for the passenger; nor can the "cord" communication be regarded as in all cases affording a means for doing so.

Having in view the primary object of the communication —the prevention of outrage—it is evident that the electric form of communication, being accessible within the compartment, acts as a more powerful deterrent than could the "cord," however reliable it might be. But although in many instances it has proved useful, it is clear that to render it available under all circumstances even greater facilities for ready access to it than are at present afforded are requisite.

How this can be assured is a question somewhat difficult to answer. Whatever the means are for giving the alarm, to be of that ready service needed it must be within the reach of a seated passenger. So arranged it would be within the reach of a child, with the natural consequence that it would be used when the need for doing so was not present.

The present mode of communication is undoubtedly capable of improvement. In utility the "cord" system is surpassed by the electrical; but whether it is not possible to produce something even more accessible than the latter has yet to be seen. Doubtless if the demand should arise, a satisfactory means of meeting it would be found.

Whatever may be done should be such as would secure the approval of all the railway companies, so that, with interchange of stock, the system on any one line may be available for use on any other line, and in connection with the stock of any other company.

CHAPTER XIII.

ADMINISTRATION.

ENGINEERING BRANCH.

IN considering the administration of the electrical department of a line of railway, economy and other—probably still more powerful—reasons will at once suggest the propriety of vesting it in the hands of one man, that by one mind all branches of the department may be so directed and dovetailed together as to operate to the advantage of employer and employé alike. Division, whether between the engineering and the traffic (message) branch of the telegraph, or between the telegraph and that which is becoming a very powerful factor, and is destined to become still more so—the electric lighting and power branch—means, at a future date, disunion, a want of unity of mind, action, and possibly interest.

Electricity is the same agent, is governed by the same laws, whether applied in the manipulation of a telegraph instrument or in the action of a motor. For those purposes which have hitherto appertained to what is termed the telegraph branch, that is the development and use of very minor electrical currents, it has been found convenient to evolve the energy needed by chemical action—the Leclanché, the Bichromate, or other form of battery. Such a battery, if of the necessary proportions, would generate the energy required for electric lighting or for the propulsion of an electric motor; but it would be a much more expensive and a much more wasteful means of doing so than is the employment of dynamic apparatus. The galvanic battery of to-day might at any moment be superseded by the dynamo, and already there are instances

to Railway Working. 261

where the current so evolved has been applied to the requirements of telegraphy. As power for the operation of motors becomes more readily obtainable, we may reasonably look forward to the replacement of the cumbrous, laborious, and costly battery, the product of chemical action now so largely in use at our important telegraph stations, by miniature dynamos which will occupy no more space than that occupied by a telegraph instrument.

It is true that the evolution of dynamical electricity has expanded the field of electrical science, and consequently that of the electrical engineer. In doing so it has also called into requisition a closer application of the laws of the electric current. In its application to telegraphy we call into use—in comparison with those large bodies necessary for lighting and power—minute quantities : quantities such that any misapplication which might arise could effect little harm. With the evolution and the employment of larger bodies it is different—mischief may arise ; but the basis is the same in each instance, and there can be no *advantage* in dissociating them. Occasions may arise where, from some purely exceptional cause, that dissociation may intervene, but there can be no reasonable doubt that the electrical department of a railway should be of the same comprehensive character as that of other departments, and that it should embrace all matters electrical.

Whether the senior post should be filled by one whose chief work has been with the dynamo or in the field of railway telegraphs, will possibly to some extent depend upon local circumstances. Electric lighting and power, apart from the principles involved, calls for certain special application. It must form a special branch of the department, but whether in the future it will become the dominating branch is very doubtful. Dynamic electricity will be employed to produce certain results. These results will be tangible ; they will be measurable in the value of their results by their cost. Here we have a plant which has cost so much money, which costs so much for up-keep and so much for working, and we get certain results from it. We can tell what those results cost, at per horse-power, candle-power, or other mode of measurement.

It is not so with a railway telegraph service. It is a necessity. It regulates and protects the traffic. Presumably prevents large expenditures which, in its absence, might arise. Its value is most apparent when it ceases to exist. It is a negative appreciation. The smoother it works the more valuable it is, yet the less is there to demonstrate its worth. Still, there can be no question of the immense value which attaches to a well established and well conducted telegraph service. In both its application and its control it calls for a wide range of knowledge, not only of the apparatus employed and the mode of applying such to the needs of the service, but incorporated with it there must be that general knowledge of railway work which will enable the application of those parts to harmonise with those wants. The man who is purely an electrical engineer for light and power purposes has much more to learn before he can efficiently and economically manage a railway telegraph service, than has a railway telegraph engineer in order to fulfil a similar part in relation to dynamic electricity.

The control of such a department must then, necessarily, be vested in an engineering rather than a commercial man. As a rule this is done, and he is regarded as the "Engineer and Superintendent" of the department. In point of fact, such is the technical character of the duties devolving upon him that it is necessary he should, to a very large extent, possess those qualities which will enable him to act, in the interest of his employers, as a manager of all that devolves upon the department.

Associated with, and subordinate to him, will necessarily be officers responsible for the efficient discharge of duties deputed to them. What these duties shall be, their extent and importance, are questions which can only be answered by a knowledge of the circumstances associated with the work. Whether the requirements of the service will extend to the appointment of superintendents of branches, or districts, or whether the need may be efficiently met by the services of an inspector are points determinable by the responsible officer alone.

In determining the location of the head-quarters of a local officer, regard should be had to means for communication with, and for visiting all parts of his section, in order that he may exercise that constant and ready supervision and control of all works, men and material placed under his charge necessary to a satisfactory discharge of his duties. His head-quarters station should, in other respects, be as nearly central to his district or section as is possible.

The extent of his territory will also be determined very much by the facilities which he may possess for ready movement from place to place.

The Construction Work—the erection and renewal of poles, wires, &c.—is at times placed under the charge of an inspector who has also the maintenance of instruments, batteries, &c., in his hands ; but more usually it is placed in the hands of a construction officer. It is, of course, preferable that the maintenance officer, who is responsible for the proper maintenance and upholding of the line, and who is therefore most intimately interested in the manner in which the construction and renewal work is carried out, should have this work under his charge ; but unless he has a very large section, or the section is a very heavily wired one, it is doubtful if it will afford full employment throughout the year for an efficient gang of men. It is not desirable the maintenance inspector's section should be so large as to preclude him from reaching its limits and returning home within the day. This will, in many cases, so determine its extent as to render the retention of an efficient construction gang upon it unprofitable.

Nor is it a good plan to provide what may be regarded as an itinerant gang—a gang to be moved from place to place under different inspectors—for the reason that all may not follow the same mode in dealing with men or material. The principles upon which work has to be dealt with are, of course, laid down by the superintendent, but this does not, nor can it, embody every detail. In carrying out these details differences may arise, not to the advantage of the work or the men. Therefore each construction gang should work under the direction of its own officer, who will know the temper of the men and their abilities, and with whose mode of dealing with the work they will also be acquainted.

Each maintenance inspector will have under him linemen in charge of certain *lengths* of line, whose head-quarters should be selected upon that basis which has already been indicated with regard to the head-quarters of the inspector.

The extent of the lineman's "length" will depend upon the facilities he may possess for visiting the various block posts, stations, and other points in his area at which telegraph and block apparatus is placed, and upon the amount of work he has to deal with at such points.

The lineman is generally provided with an *assistant*, or *battery man*. Advantage attends this arrangement. If the lineman is engaged in one quarter and his presence is needed elsewhere, the assistant will probably be able to deal with the additional demand. It is further a means of educating and bringing forward juniors for the more important post. Wherever the work to be dealt with will admit of the services of two men, it will be found advisable to station a lineman and an assistant, and unless there are obstacles in the way the lineman's "length" should be so arranged as to admit of this provision being carried out.

The strength and number of construction gangs will, necessarily, depend upon the extent of the system. Eight to ten men, according to the description of work in hand, will form a convenient number for any man to handle. If the system is large, it will be found convenient to locate gangs at two or three or more points. These points will be their head-quarters, but they will not, as such, be the point to which they will be required to return each night. Where the work in hand is situated at such a distance from their head-quarters station as to render it more economical for the men to lodge out, provision for doing so should be made. In that case they will leave their head-quarters on the Monday morning and return home on the Saturday.

The construction gang usually deals with the outdoor work, the leading-in wires, and provision of *earth* plates or connections.

The installation of instruments and batteries should be effected by the lineman under the direction of the maintenance inspector.

The maintenance inspector should be informed of all construction or renewal work ordered to be carried out in his section. The reason for this will be obvious. The maintenance inspector is the officer immediately responsible, under the direction of the superintendent, for the efficient working and upholding of the line. To enable him to satisfactorily carry out this duty, he should not only know all that has been authorised but be in a position to see that it is well done, and direct attention to any deficiency or neglect which may come under his notice. Renewal works of course come under the head of construction.

The duty of a maintenance inspector may be broadly stated as follows :—To carry out, or to see carried out, in an efficient manner all works applicable to his section, whether of construction, renewal, or maintenance, unless specially instructed to the contrary ; and, with respect to the pure and simple maintenance of his section, to see that the poles, stays, wires, insulators, instruments and batteries are well and efficiently upheld ; that failures or defects affecting the working capacity of his circuits and apparatus are removed with dispatch, their cause investigated, and their recurrence as far as possible provided against.

Every inspector and lineman should know the position which each wire occupies on the poles, and the arrangement of all circuits under his charge. A pole diagram book, a specimen page of which is shown in Fig. 141, and which should be carefully corrected from time to time, will be found a capital aid.

An important part of the inspector's duty is the careful supervision of the work to be performed by the linemen under his charge ; to see that their store sheds are in order ; tools clean and tidily kept ; that they are properly provided with stores and spare apparatus; and that their diary and stores books are duly entered up to date.

In dealing with construction work it should be the duty of the inspector to see that the works entrusted to him, or placed under his inspection, are efficiently and economically carried out; that the workmen composing the several gangs under his direction are judiciously handled; that the foreman of each gang is properly provided with stores well in advance of the work in hand; that they are properly used, that there is no waste, and that all stores recovered or surplus from any work are carefully collected and disposed of as may be directed. To see that the time books are regularly and correctly kept; that all tools, ladders, &c., are in good and trustworthy order, and available for use at any time.

Inspectors sections distant, say, one hundred miles from the departmental headquarters, may with advantage be provided with a sectional store depôt; and all linemen should possess at their headquarters a small store shed for the accommodation of such stores and apparatus as may be required by them in the maintenance of their length.

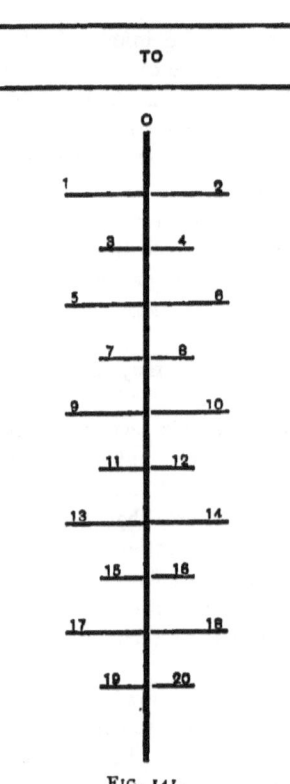

FIG. 141.

It is an excellent plan to provide each depôt, or lineman's shed, with a certain stock of stores—a " Normal Stock "—to be used as required. These stores as used should be entered by the lineman or inspector in his store-book, the book sent to head-quarters periodically—monthly or quarterly—where the items used should be summed up and entered out to maintenance, and a corresponding fresh stock returned with the store-book. The stock can then at any time be verified. The "Normal Stock" allowed the lineman being entered on the inside of the cover of the stock-book, the stock he has in hand, plus that entered out (which has not been replaced) should agree with that officially appointed as his stock.

Various systems of dealing with the stores employed in construction work exist. With some companies the stores required for a given work are requisitioned from the General Storekeeper, and debited by that officer direct to the order for the work. With others the electrical superintendent maintains such a stock of material as he feels will enable him without delay to meet the requirements of the works with which he has to deal. In that case it remains for him to account for their application.

The former method may, where a stock of material is not kept on hand by the General Storekeeper, lead to delay in carrying out work; and it is scarcely to be expected that he can determine, from time to time, the possible demands of any one department. It, on the other hand, saves labour in book-keeping by the consuming branch. Where despatch in dealing with work is, however, a necessity, it is an impossible system.

If a stock of stores has to be kept, the proper judge of what should be kept, and the proper custodian, is the engineer and superintendent of the department. To what extent the issue and consumption of these stores should be checked or recorded is mainly a question of cost. The following system is one which ensures a complete check, and is no doubt good from a disciplinary point of view.

(i) When a work is authorised, a Works Order detailing shortly what has to be done should be issued to the officers interested in its execution.

(ii) All charges, whether for labour, stores or other purposes, should be asked for under, and charged to the order authorising the work ; and all stores recovered from it should in a similar manner be credited to it.

This will necessarily entail the establishment and use of the following forms and books :—

Estimate.	Stores Issued.	Stock
Works Order.	„ Received.	Account
Requisition.	Stock List.	books.
Stores Delivery Note.		

The estimate form should provide for money expenditure for labour or other causes, and for the stores which will be required for carrying out the work, together with those which may be recovered.

The works order form should be so arranged that the officer deputed to deal with the work may, by its aid, furnish all details necessary for carrying forward the records of the department, as well as for checking the stores, &c. Specimens of these forms, giving their headings only, will be found in the Appendix. A careful consideration of them will render their object clear to the reader.

Under such a system the charges applicable to each work, whether construction or maintenance, can at all times be ascertained—the mileage and the cost for both labour and material ; and as the examining officer is aware of the material which should serve, say for the erection of a mile of work, wire or poles, an excellent check can be exercised over those to whom the material is issued ; ensuring on their part care in its administration, and the collection and return of all stores remaining unused, or which may have been recovered from the work.

An efficient maintenance will necessarily call for constant and careful attention on the part of all concerned : not only must the lineman be careful in observing the duties entrusted to him, but there should be a periodical inspection of his work and the condition of all apparatus under his charge. This periodical inspection should extend annually to the poles, wires, stays, terminations, &c. Returns of these inspections, accompanied by schedules of the work required for renewals during the succeeding year, will be found exceedingly serviceable to both the local and the controlling officer.

Returns should be rendered weekly by the linemen and others whom the chief of the department may indicate, in which should be recorded the signal boxes visited during each day, and sections of line walked and examined, or the manner in which each lineman or assistant lineman is otherwise occupied.

The method and scale of pay, the classification of the men, the sum to be allowed men if required to lodge out when working at such a distance from their head-quarters as to render their return each night uneconomical, and other kindred questions, is matter for the determination of the chief, and for regulation somewhat in accord with the custom and other ruling factors of the locality. That harmony should in these matters exist also between the employing companies, and between the employer and the employed, needs no exposition at my hands.

All important wires should be tested for insulation daily, or if the number to be dealt with is too numerous to admit of this, one portion should be tested one day and the other the following day. It is usual to effect this in the morning between 7.30 and 8 o'clock, or thereabout. Where a wire shows unusual loss, or has fallen below the standard, the position of the defect should be localised without delay— within the time allotted if possible, otherwise as soon as possible thereafter.

The testing stations should be located within 100 miles of each other, and provision for disconnecting, earthing, or otherwise dealing with the wires for the purpose of localisation, should be made at intermediate points as may be convenient. A lineman's station is a good point, as he can then be sent on either side of his station, according to the locality in which the fault has been ascertained by the test clerk to exist.

Most British railways are maintainers of wires for the British Postal Telegraph Department. Many of these wires serve most important circuits, as for instance, between London and Glasgow, Edinburgh, Belfast, as well as the lesser distances between the metropolis and Birmingham, Manchester, Liverpool, &c. A certain system of testing is in operation by the Post Office. It is desirable that employed by the railway company should conform thereto: comparisons may then readily be made in respect of the results obtained by either party.

Every fault which arises and is removed by the lineman, whether on a telegraph message circuit or on block or repeater wires, should be reported upon a form provided for the purpose by the lineman as soon as possible after its removal. His report should be duplicated, one copy being sent to the superintendent, and one to the local inspector so that that officer may be acquainted with its cause, and take such steps as he may be able to prevent a recurrence—a course which will also naturally be followed by the chief officer.

Electric Light and Power.—There is of course no reason, where generating plant is established for electric lighting or power purposes, why it should not come under the supervision of a senior local officer, such as the superintendent in charge of the telegraphic service of the district, provided he possesses the needful qualifications for the exercise of such supervision. The laws which govern the application of electricity to the purposes of lighting or power are, as has already been mentioned, equally applicable to its application to telegraphy: the one is generated by motion, the other by a primary battery. The former is required and is produced in large bodies, and its use is confined to spaces within as small a radius of the generating centre as circumstances will admit.

In the employment of these larger bodies of energy we have to deal with that which, if not properly controlled and safe-guarded, may lead, as with steam and other powerful agents, to undesirable consequences. This is not, as a rule, the case with energy evolved from a primary battery, or such a form of battery as is employed for railway signalling and telegraphic purposes. Hence, whereas with the one there is not that power of doing damage, and consequently not that great need for exactitude, with the other this is imperative.

In telegraphy and in signalling apparatus we have to deal with resistances seldom embodying less than 100 ω. With electric lighting the circuit resistances are measured by the decimal parts of an ohm.

A dynamic electrical generating plant is a plant which, however, requires constant attention. Its condition must always be good. When in operation it requires constant supervision, just as does a locomotive engine. This supervision necessitates the entire services of a staff consisting of one man upwards, according to the magnitude of the plant. Whether it entails the service of one man or of many men, the officer responsible for its good running must be there, and hence, although such generating plant may be supervised by the divisional engineering officer, the immediate responsibility for and management of it must rest with the chief man in charge of the generating station.

In sufficiently large installations a local or resident engineer is desirable, and he, under the chief of the department or a district engineer, should be amenable for the efficient working of his station. Daily reports or returns of the working of the plant under his charge, of any failures of arc lamps, or other items which it is desirable the divisional engineer or superintendent of the department should be acquainted with, should be punctually rendered.

As the electrical department must naturally be responsible for all matters electrical, so should the fullest power for dealing with the apparatus placed under the charge of the department be accorded it. It may be that the department may not, in its incipient stage, or even at a later period, possess the means of dealing with all those repairs, mechanical or electrical, which may be called for; but where these matters can be dealt with by other departments of the service, the resources of such departments should be at the disposal of the officer controlling the electrical department.

During the summer months the demand for artificial light is considerably less than during the winter. A well organised staff cannot always be re-formed when once disturbed. It is well, therefore, where such exists, not to dispense with members of the staff during the shorter nights. Work can usually be found for them. Each man will require annual holiday relief. Lamps will need careful overhauling, re-painting, &c., as will also lamp pillars and fittings, together with many other matters both in and out of the engine room. The summer months is the period for this and for thoroughly overhauling everything. Not only does it find employment for, and so keep together a well matured staff, but it fixes the responsibility upon those who are locally amenable for the good working of the installation; there is no division of responsibility. Division of responsibility is the *bête noir* of success.

TRAFFIC BRANCH.

Circuit arrangements.—The telegraph system of a railway service will naturally centralise—
(i) At the head-quarters station of the system.
(ii) At its sub-centres.

In dealing with the chief centre of the system we have thus two classes of circuits, one affording communication with the sub-centres or chief points of the service, and another embracing the near or local stations.

The extent of the local circuits, the radius they shall cover, or the number of stations each may serve, are points to be determined by local conditions, the probable number of messages required to be dealt with, and the means for dealing with those messages, whether at the hands of a signalman or those of experienced telegraphists.

Retransmission of messages means delay, the cost of clerical service and office accommodation. Instances will arise where it is desirable that a circuit shall be extended to head-quarters. Where this is the case, we have to consider the annual interest on the cost of the extension of the circuit affected, plus the cost of its annual up-keep, as against the cost of clerical aid, &c., involved in the retransmission of the messages. Another and equally important point for consideration is the power of the circuit over which the work

to Railway Working. 273

has to be retransmitted to carry the same. If the circuit is already fully loaded, then the circuit calling for modification must be carried forward to the head-quarters station, or to some other point which can equally well serve it.

The character of instrument which should be employed will depend largely upon the amount of work to be disposed of, and in no small measure upon those who have to deal with it. The circuits communicating with the chief points will in all probability be worked entirely by experienced manipulators ; and in this case the bell instrument previously referred to, or the sounder, will doubtless be found most suitable. For the local circuits communicating with signal boxes, largely worked by signalmen, the single-needle type of instrument will be found the most simple and serviceable. Throughout the railways of Great Britain this class of instrument is in general use for the purpose, and is well understood by the signalmen. Where fixed in signal cabins, it should be provided with the vertical handle in preference to the pedal form of commutator. At telegraph offices the pedal form may be used, and as has already been pointed out, the bell form of instrument may with advantage take the place of the single-needle at those telegraph offices which form the terminal or transmitting station of the circuit.

What may be termed "through" circuits—those communicating with distant points—necessarily require that a limitation should be placed upon the number of stations which they shall serve. It is undesirable the number should exceed six ; and it is perhaps needless to point out that such circuits should not be occupied by local work—messages between adjoining or near stations. All messages, whether passing between stations far apart or near together, occupy the wire to the prejudice of other work waiting its turn ; and where local communications are frequent, it is desirable to consider the propriety of providing a local circuit for its accommodation rather than allow such to occupy a busily engaged "through" circuit.

Whether the number of messages passing between any two points will warrant the establishment of " Duplex " or " Quad-

T

ruplex" working will naturally make itself apparent. It will be manifest that to economically work duplex the messages must be fairly distributed, i.e. they should flow equally both ways—up and down. If the work is all in one direction "duplex" is of no service.

The employment of the "Wheatstone," or that of the "Multiplex" type of telegraphy, has not, so far, been employed on railway systems; nor is it probable, having regard to the very partial demand which is likely to arise at even the chief centre of a large railway service, that the demand for such will for some time become at all pronounced.

Reorganisation of Circuits.—However well a telegraph message system may be organised, it will from time to time call for rearrangement. The exigencies of railway service are ever varying. It is the traffic which gives rise to the demand for telegraphic communication. The traffic itself depends upon the trade of the locality. As this varies so will the traffic, and so will the demand on the telegraph wires. The number of messages passing over wires devoted to commercial messages affords a very fair index of the business of the locality or of the country. To some extent the same may be said of the telegraph message work of a railway system. There will, of course, arise exceptions. Congestion of traffic emanating from climatic conditions or other causes, will lead to increased telegraphic correspondence. This may lead to a block on the telegraph wires. Experience of local conditions and causes must prove our guide in what manner such demands shall be provided for. If the congestion is of frequent occurrence, clearly a remedy is needed. The circuits must be reorganised or supplemented. The work dealt with by each station upon the circuit affected should be analysed in order to learn—

(*a*) The number of messages dealt with each hour.

(*b*) The delay which each message has experienced.

(*c*) The station to which each message is sent.

A return for the hours during which congestion occurs should be rendered by each station daily, until it is felt that the normal condition of the work passing over the circuit has been ascertained.

A careful consideration of these data will show between what points the chief of the work arises, and readily enable the responsible officer to determine in what way the difficulty should be met.

The form shewn on p. 276 will be useful for recording the delay, and number of messages dealt with per hour. It may be used not only for this purpose, but as a daily record of the work disposed of by terminal stations on heavily worked wires. The messages are entered upon it in their respective columns, according to the delay which they have experienced, by the clerk as he disposes of them. The number of messages *received* is only material in enabling the return to show the total number of messages dealt with per hour. If it is desired to know what delay attaches to them, similar returns must be kept at those stations from which the messages are received.

Unnecessary use of Wires.—Those who are accustomed to telegraphic correspondence, and especially those who have an intimate acquaintance with a free use of the telegraph wires, will know well what an inducement the telegraph message is to dispose of correspondence which otherwise might equally well, for the purpose in view, have been disposed of by a letter sent by train or by post. It is so much easier to write a message than to write a letter. When we write a letter the conventionalities of society demand a degree of courtesy which is readily excused, and which would seem absurd, in a telegram. It is not quite clear why it should be so ; why a letter—especially a business letter—should not be just as acceptable if couched in the same concise terms considered so proper for telegraphic correspondence. However this may be, there is no doubt the telegraph, especially where its use entails no cost upon the user, is frequently used when a letter would answer the purpose. This should be judiciously checked. It is not desirable to limit the use of the wires so as to impede or to jeopardise traffic, but every flagrant abuse of them should be recognised.

Use of Prefixes.—Even where it is necessary to revert to the use of the wires, there are many messages which are, although of importance, yet not of that importance which

RAILWAY.—TELEGRAPH DEPARTMENT.

Circuit _____ . Station _____ . Date _____

TABLET CHECK.

Hours.	NO. OF MESSAGES SENT WITH DELAY UNDER							RECEIVED	REMARKS.
from to	5 min.	10 min.	15 min.	20 min.	25 min.	30 min.	Over 30 min.		
A.M.									
10 to 11									
11 ,, 12									
P.M.									
12 to 1									
1 ,, 2									
2 ,, 3									
3 ,, 4									
4 ,, 5									
5 ,, 6									
6 ,, 7									
Totals .									

Total Forwarded _____ . Total Received _____ . Signature _____ should entitle them to precedence of others. In all systems it will be found desirable to establish an order of precedence to be observed by all who have to deal with the transmission of messages. The following is that observed by one of the largest of the British railway companies.

PREFIXES INDICATING ORDER OF PRECEDENCE.

D N G —Danger Signal.
G M —Extremely urgent Message.
M T —Train report.
T A S —Train Message, for delivery.
T A X —Ditto ditto for transmission.
R S P —Engine Driver's and Guard's Relief Message, for delivery.
R X P —Engine Driver's and Guard's Relief Message, for transmission.
S P —Special Service Telegram, for delivery.
X P —Ditto ditto for transmission.
D S —Telegraph Engineering Message, for delivery.
D X —Ditto ditto for transmission.
S G —Repetition and Urgent Telegraph Service Message, for delivery.
X G —Ditto ditto for transmission.
D B —General Railway Message, for delivery.
D L —Ditto ditto for transmission.

The D N G prefix is established, and especially reserved, for use in cases of great urgency affecting the safety of the traffic. It should have the power to take possession of the circuit at any instant, no matter how occupied, unless, of course, it should happen to be engaged with a message of the same character and bearing the same prefix. In all other cases it is desirable each message should be preceded by a time code, the number of words, and the address from and to. The urgency which calls for the use of this prefix dispenses with everything except that required to accomplish the object of the communication. Hence the interposition of the signal D N G on the circuit should be sufficient to stop everything. The station or signal-post using it should obtain as soon as possible the attention of the station required, give the code of his own station, and proceed with the required announcement, as for instance—

D N G. Stop all up (or down) trains.

Having accomplished this, a further message, under a less urgent prefix, explaining the occasion for the D N G should be sent.

A strict check should be kept upon the use of this prefix. A copy of the communication, with a full report thereon, should be sent to the Superintendent of the Line, as well as to the Superintendent of the Telegraph Department and such other officers as may be desired. The same may be said with respect to the use of the prefix G M. It is an important prefix, and if to be of service must not be employed to the detriment of other messages unless there is good reason for doing so.

Train reporting will be referred to later on. R S P is a prefix provided to ensure the speedy transmission of a class of messages only recently found to be necessary. Where traffic becomes congested, delay to certain trains, those of the least importance in respect of speed, must arise. It is necessary in such cases that a message should be sent by the officials in charge of such delayed trains, to those points from which relief may be sent them on completion of their appointed hours of duty.

The attachment of the prefix should be the duty of the station master, or where telegraph offices are established, of the clerk in charge of that office.

Code Time.—Subject to the precedence accorded by the prefix, the order of despatch of all messages is further subject to its code time. Messages bearing prefix D N G do not require code time, and those sent as D S or S G seldom have the number of words comprised within the message signalled. All other messages, however, should bear the code time at which they are handed in. This code time is obtained by applying the letters of the alphabet from A to M (excluding the letter J) to the diurnal hours, and the letters R S W X to the minutes intervening between the hour letters, as illustrated in Fig. 142. Thus the time 7.12 (twelve minutes past seven) would be represented by G B S, 7.15 by G C, and 7 o'clock by G.

The signalling of the code time is usually followed by the number of words—that is, the total number contained in the message itself and the address from and to, in order to guard against omission on the part of either the sending or receiving telegraphist.

FIG. 142.

Telephony.—Although long distance telephony has so largely entered into commercial life, it cannot be said to have established a hold upon British railway systems. Telephonic communication between signal boxes, and between stations and signal boxes, as also between the various offices of the head-quarters of the company, and again locally between the chief offices of the most important centres has become necessary, but such applications are entirely local and extend to no material distance. Nor, probably, is there that need for direct and personal communication between distant points of a railway system that appertains to a commercial transaction. The correspondence which on a railway has to be carried on by the aid of the telegraph, as a rule requires to be remitted to paper for future reference if necessary, and hence, whether made by telegraph or telephone, has to pass through other hands than those with whom the communications emanate, and accordingly has to be written down.

It is generally conceded that for signal-box telephonic communication the circuit arrangements should limit the number of instruments to ten. That is, that this number should in no case be exceeded. It may, of course, be reduced to meet whatever will comply with the need of the case.

Where several instruments are arranged upon the same circuit, it will probably be found convenient to employ the letters of the Morse alphabet for the call signals of the respective boxes or stations—a short ring representing the dot and a relatively long ring the dash. The duration of the "dash" should be three times that of the "dot." The duration of the "dot" may be as short as is compatible with the production of a ring on the bell.

Collection and Examination of Messages, &c.—The telegraphic correspondence of most railway companies has now assumed such dimensions as to render the possibility of collecting and comparing the forms relating to each message a gigantic task—one involving so large an amount of labour, and becoming, consequently, so costly that it has been perforce abandoned. Still it is desirable the message forms and train report sheets for each day should be, by each station, carefully collected and docketed, and at the end of the week sent in to the head-quarters station, to be there entered and checked. Without some system of entry and checking it will be clear neglect may pass unnoticed until a demand for some important correspondence arises to illustrate the absence of that check so much needed.

At the same time, although the magnitude of the work prevents the whole of the message correspondence of a system being dealt with in that comprehensive manner one would desire, much good may be done by taking in hand the forwarded work of such a number of stations as time and staff will admit, and comparing the delivery copies therewith. Any errors will thus be brought out, and omissions in date, code, &c., detected. It will further be useful for checking any apparent unnecessary use of the wires or employment of unnecessary words. Reference of these to the stations in fault will prove influential in the suppression of further irregularities.

Office Check Sheets.—In all busy offices a system of *checking* messages as they pass through their various stages is very desirable: it prevents delays and other irregularities. The system involves the provision of *check sheets* for the different

classes of messages. Dealing with a forwarded message :—
When first handed in, the next open number on the check
sheet is accorded it ; it is then passed to the instrument, and
when despatched, collected and returned to the check-sheet
clerk, by whom it is examined, and, if properly dealt with, its
number is ticked off in order to show that it has been checked.
The message is then filed. Received messages pass through
the same ordeal, the check required in this instance being
that the message has been duly delivered and its receipt
acknowledged by signature.

Transmitted messages require on their receipt to be
handed to the check clerk for numbering. Thence they are
referred to the instrument on which they are required to be
sent, and after this has been done are passed to the check
clerk before being put away.

In offices where the number of transmitted messages is
large, an automatic numbering stamp may with advantage
take the place of the check sheet. In this case each transmitted message is collected as received, placed in order of
time with others which may at the same time be brought to
the stamping table, and stamped in consecutive order. They
are then sent to the circuit for transmission, and on this being
accomplished are examined by the stamping clerk and
placed away in the order of their number. The stamping
clerk is of course responsible that no number is missing.

It will be noticed that the "No." columns of the following
forms are provided with unit figures 1 to 0. The object is
that the check clerk may, by preceding these units with other
figures, make use of them for any numbers required to meet
the day's registration.

TRAIN REPORTING.

Where a mixed traffic has to be conducted over the same
lines of rails, it is of material importance that the progress of
trains of a certain class should be reported as they leave certain stations, to stations and signal posts in advance for a

CHECK SHEET.

FORWARDED MESSAGES.

Station, ———— 189——

This Check Sheet is intended for the use of larger Stations only. It is to be filled in as the messages pass through the Clerk's hands, and tied up with each week's message papers, and forwarded to ———— at the end of every week.

Date.	No.	Destination.	Date.	No.	Destination.	Date.	No.	Destination.
	1			1			1	
	2			2			2	
	3			3			3	
	4			4			4	
	5			5			5	
	6			6			6	
	7			7			7	
	8			8			8	
	9			9			9	
	0			0			0	

CHECK SHEET.

RECEIVED MESSAGES.

_____ Station, _____ 189__

This Check Sheet is intended for the use of larger Stations only. It is to be filled in as the messages pass through the Clerk's hands, and tied up with each week's message papers, and forwarded to _____ at the end of every week.

Date.	No.	Station from	Date.	No.	Station from	Date.	No.	Station from.
	1			1			1	
	2			2			2	
	3			3			3	
	4			4			4	
	5			5			5	
	6			6			6	
	7			7			7	
	8			8			8	
	9			9			9	
	0			0			0	

CHECK SHEET.

TRANSMITTED MESSAGES.

_____ Station, _____ 189___

This Check Sheet is intended for the use of larger Stations only. It is to be filled in as the messages pass through the Clerk's hands, and tied up with each week's message papers, and forwarded to _____ at the end of every week.

Date.	No.	Office.		Date.	No.	Office.		Date.	No.	Office.	
		From	To			From	To			From	To
	1				1				1		
	2				2				2		
	3				3				3		
	4				4				4		
	5				5				5		
	6				6				6		
	7				7				7		
	8				8				8		
	9				9				9		
	0				0				0		

given extent of line, in order that a clear route may be maintained for the more important trains.

When our main routes are provided with "fast" and "slow" lines the necessity for this will largely disappear, and the traffic on both routes will moreover be greatly facilitated, for it is scarcely possible to truly gauge the enormous loss of time which has now to be incurred—and cannot be avoided—in the stoppage and shunting of trains in order to admit of the passage of others timed to travel at a higher rate of speed. Without some method of making known to those who are responsible for the safe conduct of the traffic, the time at which they may expect and so make provision for the passage of these more important trains, clearly it would be impossible for them to do so unless it could be ensured that every train should keep absolutely its appointed time.

The establishment of such a system of signalling involves the division of the line into convenient lengths, and the provision, in connection with each section, of a telegraph station for the purpose of signalling to those points at which the information is desired, the time of departure of the trains required to be reported. The length of each section will be governed by the requirements. It must not be too long, or any irregularity which may occur in the speed of the train after it has left the reporting station will vitiate the value of the information; and if it is too short, the time between the departure of the train and its passing will be insufficient to provide for its progress without checking it. Twenty to thirty miles will probably be found a convenient length.

In order to provide sufficiently early intimation to the near stations on each section, it is desirable the reports applicable to the first section should also be signalled forward to certain stations belonging to the next section, and to such chief points beyond as the exigencies of the traffic demand. The object is to get information of the running of the train so as to make provision for it with as little sacrifice of time and speed to other trains as is possible, and to do this without multiplying reports and overcrowding the wires with work.

Thus, assume a section of line 90 miles in length divided

into three sections—say A B, B C, C D. On the train leaving A its departure would be signalled to the posts indicated by the instructions between A and B, and to such other points beyond B as might be indicated. On the train passing B it would be signalled through the B C section, and to such points beyond C as were indicated.

The matter signalled should be condensed into as small a space as possible. Each train is known by its number; i.e. the number under which it is shown in the "Working Time Table"; or if it is a special, by the number it bears in the "Special Time Table"; and thus it would be signalled in the former case as—

MT 69 at 5.42,

which interpreted would mean "Working Time Table train No. 69 left this station—that is, the station from which the train is being signalled—at 5.42"; and in the latter—

MT special 130 at 10.5.

The order of precedence of train reports must necessarily be that of the time announced in the report.

In the majority of instances this information is required by the signalmen at those posts at which there exists siding or other accommodation for shunting trains. It is also of importance to the signal box staff at junctions and the chief stations. At the latter points the information is equally of service to the platform staff, and provision to meet this is necessary.

Where the telegraph signalling has to be dealt with apart from the signal box, as is usually the case at all important stations, telephonic communication is established between the signal box and the telegraph office. By its means the departure of the train is announced. Where this provision is not made it is usual for a member of the platform staff to report the time of departure to the telegraph office. The practice pursued by the author is as follows.

In either case the necessary details are entered in a book by the telegraph clerk, who has to receive the report and circulate it. If received from a signal-box, a number or a letter is recorded as a check that the report has been duly received. The following *train slip* is then filled in and sent to the instrument for distribution of the information to the stations named in the first column.

_____ RAILWAY.—TELEGRAPH DEPARTMENT.

Station, _____ 189

No. of Train _____ *Station* _____ *Time of Dep.* _____

Stations To.	Time Reported.	Circuit.	Signature.

Each post on receipt records the same on a form of the following character, the other necessary details having been previously filled in. This sheet serves for reference for any of the trains throughout the day, and is purely for the signalman's use.

_____ RAILWAY.

_____ *Station*, _____ 189

_____ *Circuit*.

Station.	No.	Due.	Left.	Late.	Time Reported.	Signature and Remarks.

288   *The Application of Electricity*

For the information of the platform staff another form is filled in and exhibited on a suitable board, usually arranged in one of the office windows, accessible to those who require from time to time to consult it. This form, as will be gathered from a perusal of the following, provides for details of the several reports for a considerable section of the line—in the instance in point between Birmingham and Derby—i.e. it affords the necessary data from Birmingham, Tamworth and Burton, the three train-reporting stations for the Birmingham-Derby section of the Midland company's system.

MIDLAND RAILWAY.

DERBY_____day of _____189

No.	Description.	Birmingham.			Tamworth.			Burton.		
		Due.	Left.	Late	Due.	Left.	Late.	Due.	Left.	Late.

It is the duty of the telegraph department to issue instructions from time to time as to the trains which shall be reported and the stations to which the reports are to be sent. Each reporting office should retain in a conspicuous position, for the information of the staff of the office, a list of the trains to be regularly or specially reported, together with other necessary data.

Referring once more to the list of prefixes, it will be observed that provision is made for a T A S (*train message*) report. Such a prefix will be found useful for reporting trains out of course—trains the reporting of which has not been provided for, but which, owing to pressure of traffic or other causes, it is thought desirable should be reported specially.

To ensure prompt despatch, it is necessary that precedence should be assured. Hence the provision of a special prefix for the purpose.

The varying requirements of a train service, increased facilities for its disposal, &c., render a periodical reconsideration of all standing orders for the reporting of trains desirable, otherwise in course of time many will become obsolete and occupy the wires to no purpose.

TELEGRAPH MESSAGE CODE.

With some railway companies a telegraph code representative of terms and sentences in general use is in operation. The London and North Western employ a letter code. Certain other companies a code based on the names of birds, beasts, fishes, &c., the varying classes being appropriated to the different departments of the service. Thus, for instance, the names of beasts and birds being made applicable to terms in general use for goods and mineral traffic, a message received quoting the name of a beast or a bird would be at once understood to apply to that department.

There can be no question that the employment of a code will prove of advantage in reducing the time which the subject of the message would otherwise, if transmitted in full, occupy in passing over the wires. In other words, a larger number of messages written in code can be sent over the same wire in the same time than if those messages were signalled in full; but in considering this question we have to consider the effect the use of a code will have upon the object of the communication, regarded in the light of a telegram. The essence of the telegraph is despatch. A code message will take little if any more time in transmission, than an ordinary message, word for word. Any material loss of time will arise in deciphering the message. It is urged by those who have used the code that in a very short time those who are deputed to deal with messages of this character become so accustomed to the code that they find it unnecessary to refer

U

to the code book. Cases have, on the other hand, been cited where, owing to the person to whom such a message has been delivered not having the code book with him, he has been unable to deal with it.

A code kept within a certain limit—say not exceeding 300 words—may, having regard to the fact that a proportion only of this number will apply to any one department, and that such messages will, as a rule, be dealt with by the same members of the staff of that branch, possibly be found serviceable and fairly expeditious. Still, delay in dealing with such messages will under any circumstances occasionally arise. It would be unreasonable to expect that the class of men to whom such messages have to be entrusted will, in all cases, be able to deal with them without reference to the code, and where this is the case there cannot be the same ready despatch as is practicable with messages the text of which is written in full.

Where the code is of a more extended character the delay must of course be intensified. In the majority of cases reference to the code will be necessary. At the same time it may be argued that the messages are not of that urgent character which will not admit of the delay arising from reference to the code book. If this is so, then no harm can be done beyond the possibility of error in transmission or misinterpretation on the part of the person who has to deal with the communication.

Broadly it may be conceded that—

(i) Where a telegraph system is capable of dealing with the work entrusted to it *in extenso*, it is preferable that it should do so rather than that a code—even of a limited character—should be employed.

(ii) Where a code is required, that a word code is preferable to a letter code.

(iii) A limited code is preferable to a more extended one.

(iv) The employment of a code is, so far as the railway telegraph system is affected, serviceable only in the saving of time effected in despatching such messages over the wires. Where the code is available for messages which have to be paid for, there may be a saving commensurate with the number of words which the code represents.

(v) Where code messages require to be translated before distribution by the telegraph office, it is doubtful if any economy attends the employment of the system.

The application of a code to so large a telegraphic correspondence as accompanies railway working when conducted over the company's own wires, is a very different thing to a fairly full correspondence conducted by a commercial house by means of messages which have to be paid for at the ordinary tariff rate. In the one instance the cost will not exceed one penny per message; in the latter, based upon the present tariff for messages over the British postal system, the cost would for the same message amount to, at the least, ten times this sum.

A letter code is of course equally as representative as a word code, but it is not, where popular words are employed, so easily remembered; and it is, telegraphically, more liable to error than words. Thus the letters D T (D— - - T—), which may be assumed to represent any sentence, are capable, unless carefully spaced when signalled upon the telegraph instrument, of being transformed into the letter X (— - - —), or the letters T I T (T— I - - T—). It is therefore very desirable when framing a letter code, to select such letters as are as little capable as possible, should the elements of which they are composed be irregularly or carelessly signalled, of being transformed into the characters which represent other letters, and which of course may be indicative of a very different meaning.

A very limited letter code is used by most railway companies for rolling stock purposes. Information is collected each weekday morning from outlying stations or districts of the stock on hand, so that it may be disposed of to meet the day's requirements with the greatest advantage and the least amount of haulage, wear and tear. These telegraphic returns are similar in character day by day, and instead of signalling in full "Horse-box," the initial letters H B are used; and a similar course is pursued in announcing the number of other stock on hand or required; but this, it will be observed, is by no means so comprehensive as a code applicable to sentences conveying various instructions or inquiries, and the interpretation thereof is readily acquired, owing to the code being composed of the initial letters of the name of the vehicle, and not an arbitrary application of letters or words to certain expressions.

The advantages attending the employment of a code call for careful consideration prior to the adoption of any such device for the mere relief of the telegraph wires. Its introduction can scarcely effect any reduction in staff. In many instances its use will not be attended with that despatch of which an instruction or an inquiry in plain English is capable; while there are other possible undesirable consequences, the character and nature of which will readily present themselves to the consideration of a thoughtful mind. The first question which may well be asked is :—If we do not adopt this code, what is it going to cost? Its non-adoption may mean a certain increase of wire; the establishment of duplex or quadruplex working; an increase of staff! What will be the cost of this, and will the adoption of a limited code save that cost, and at the same time maintain the efficiency and value of the telegraph service?

APPENDIX.

USEFUL RULES

For the Guidance of Attendants and others engaged in Electric Lighting Duties.

It is important that every one who has any duty to perform in connection with electric lighting machinery and apparatus, should bear in mind that the lighting of those buildings, yards, &c., to which it is applied is dependent thereon, and that the apparatus employed for the purpose must, as far as is possible, at all times be ready for use.

The foreman of every installation, or in his absence, the leading-man, on the approach of fog, or darkness from other cause, should, after satisfying himself that the lamps on the circuits are ready for use, get the machinery into motion, and the lighting in operation as soon as possible.

Where the lighting of the station, buildings, &c., is entirely dependent upon the electrical machinery, steam should be in readiness for use at any moment.

All machines, wearing parts, belts, circuits, &c., should be carefully tested and examined when shut down, so that any defect may be at once rectified.

One hour prior to the usual time for starting, the attendants in charge of engines and dynamos should carefully test and examine all parts of the machinery, circuits, &c., under their charge, and satisfy themselves that all is in good working trim and capable of carrying on the run for the time required. If any doubt should arise, the machinery should be started at once, in order that its condition may be practically tested.

Should anything transpire which may prevent the lighting, immediate notice should be given to all whom the absence of the light will affect.

When necessary to change over from one set of machinery to another, or from one machine to another, every effort should be made to effect the change without interrupting the lighting.

A sufficient stock—not less than one month's supply—of oils, carbons and all other necessary stores, should be kept on hand.

The resident engineer, inspector or foreman, should be responsible for the due and efficient performance of the duties of the staff. He should appoint the duties and rounds of the dynamo attendants, trimmers, cleaners, &c.; see that those duties are faithfully carried out; that all parts of the apparatus are kept in perfect working order; take every step in his power to ensure the successful working of the machinery and apparatus entrusted to his care; and bring under the notice of his superior officer without delay any neglect of duty.

Hot carbons when removed from a lamp should be placed in a receptacle safe from fire.

Foremen and trimmers should satisfy themselves that no possibility exists of burning particles falling from or passing out of the lamps.

Open lights should not be used in goods sheds, depôts or elsewhere, without special sanction from the senior officer of the installation.

Whenever steps or ladders are required for use in public roadways, sidings or lines of railway, every precaution in order to avoid accident by or to passing vehicles or foot passengers should be adopted. If necessary, a man should be posted at the foot of the steps or ladder to protect the workman standing thereon, and warn approaching foot passengers or vehicles.

All men employed, whether trimmers or others, should be required to take every precaution to avoid the possibility of accident to themselves or others. Any one called upon to deal with apparatus through which the electric current is at the time passing, or during such time as there may be a risk of the current prevailing, should use sound india-rubber gloves.

Each trimmer, on completing his section, should report the same personally, or by telephone, where such is provided, to the engine room.

Workmen, when going off or coming on duty, should leave the premises by the route indicated by the officer in charge of the installation, and which should be that route which will afford the most proper means of reaching the public street or roadway without crossing working lines of railway.

Dynamo Room.

The dynamo room should always be clean, and everything in its place, ready for use.

Immediately a machine is shut down, it should be cleaned and prepared for further use in case required, and covered up till required.

All machines, circuits, and parts of apparatus should be tested when shut down, and an hour before being started. Where low-tension machines are kept running for several hours, the machine, as also the circuit, should be tested for earth during the run; but such tests should in all cases be made through a "resistance" approved for the purpose.

Where the frame of a high-potential machine is insulated from the bed-plate, daily tests should be made in order to ascertain if there exists any leakage from the coils or other parts of the machine to the frame from which it is insulated. If such is found to exist, those attending it should be warned personally, and a notice in writing affixed to the machine.

Before starting, all binding screws and bolts should be examined. If a binding screw is loose, its surface should be examined. If found to be burnished, a fresh surface should be provided.

A diary and a duty book should be kept at each depôt. In the former should be entered, by the chief attendant in charge, the condition of matters generally when leaving and when coming on duty, together with any circumstances which may have occurred during his duty, such as changes in the machines, irregularities of any description, details of the run, &c. &c.

Iron and steel tools should be kept clear of all machines. Metal work should not be filed in their immediate neighbourhood. A copper oil-feeder with an insulated nozzle should be employed for lubrication.

Precautionary Instructions in relation to the Avoidance of, and for the Treatment of Sufferers by, Accidental Shock from Electrical Current.

If ordinary precaution is observed in handling the electrical cables, wires or apparatus, no danger need be apprehended; but where indifference to well known restrictions, or negligence in any form, is allowed to prevail, serious consequences may ensue, not only to the individual guilty of such negligence, but to others who may be associated with him in the discharge of his duty. Regard any apparatus with which you may have to deal as capable of affording a shock, and adopt precautions accordingly.

No person should be entrusted with the performance of a duty until he is perfectly competent to deal with it.

When it is necessary for regulation or other purposes to touch parts of such apparatus when the current is active, the individual should be careful that his entire person is insulated from the earth, and from anything, other than an insulating medium, which connects with the earth. This may be best achieved by standing on an india-rubber mat. On no account should he touch parts of the machinery, cables or apparatus with both hands at the same time, unless his hands are protected by sound india-rubber gloves. Shoes with india-rubber soles may be worn as an extra precaution, but the wearer must never rely upon them for insulation. Dynamo-room men should keep such shoes in the dynamo-room, and not wear them indiscriminately in and out of doors. The soles should be examined daily in order to ascertain that they are sound and dry.

The inspector and foreman of each depôt will be held individually responsible that a sufficient supply of india-rubber gloves is kept on hand to replace broken ones, and that one or more pairs (according to the magnitude and importance of the installation) are retained in a closet provided specially for the purpose (and marked I.R. GLOVES), so as to be readily accessible at any time.

Workmen are required, after using the gloves, to return them to the officer in charge of the dynamo or distribution-room; and that officer will be held responsible that all gloves issued to workmen are returned to the closet, or otherwise accounted for by an entry in the diary.

Should a lamp require attention during the time the current is flowing, the lamp is to be switched out of circuit before being touched, by the switch at the base of the lamp pillar, where such is provided; otherwise by the switch on the lamp, which, it must be borne in mind, does not in all cases insulate the lamp.

In cases of accident where a man has become paralysed by the current and is unable to extricate himself—

(1) The first effort should be to switch off the current.
(2) Failing the power to do this, to divert the current from the sufferer.

The first instruction is easily effected where the accident arises within ready reach of the dynamo, distributing-room or switch.

The second instruction requires careful handling on the part of the rescuer. He has to bear in mind that if he touches the sufferer with his naked hands—even by his clothing if damp from rain, or the perspiration of the body—he may be placing himself in the same dangerous position as he whom he desires to rescue.

It is necessary he should keep himself insulated from the cable or apparatus, and from the sufferer, as well as from the earth.

Time is of the utmost importance, and will probably not admit of running for gloves or india-rubber mats. An instant of time may, if judiciously employed, suffice to release the sufferer, and any dry article of clothing may serve as an insulator for that short period.

If a piece of wire of sufficient capacity—an iron rod or any form of conductor—is at hand, the rescuer should, first, connect one end of the wire or rod to earth, and, failing a better means, divest himself of his jacket or coat and use it to insulate his hands from the other portion of the wire or bar and the cable whilst he attaches it to, or places it in contact with, the cable (or whatever may be the source from which the current is passing into the sufferer). The attachment should be as near to that part which is in contact with the sufferer as is possible. This may divert the current to the earth and release the sufferer.

Whether it does so or not, the rescuer should, with continued caution, proceed to remove him from the position in which he is placed. He must insulate his own person, by standing on a dry board, dry clothing, a bundle of dry straw, hay, &c. He must protect his hands by interposing as many thicknesses as possible of his dry clothing between them and the sufferer; and in this way endeavour to withdraw or raise and insulate the sufferer from contact with the earth. When insulated, the current will cease passing through him, and the cable, &c., may be withdrawn from contact with him. In doing so the rescuer must exercise great care that he does not place himself in a position to receive a shock.

Where a man is receiving a current through his person by having placed one part of his person in contact with one portion, and another part in contact with another portion of a cable, &c., thereby forming of his person a portion of the circuit, the most ready means of delivering him is by short-circuiting such portion of the apparatus immediately outside the points of contact with the sufferer.

In cases where a man has received a serious shock, and life appears extinct, efforts similar to those employed in cases of drowning should be made to restore animation. Experience has shown that the D'Arsonval system has been very successful. The treatment is as follows:—

Lay the patient on his back in the open air. Remove his neckcloth and unfasten his collar. Open his mouth, and, taking hold of the front part of the tongue with your fingers—either bare or covered by a handkerchief—very slowly draw the tongue forward, and as gently let it go back again sixteen times a minute. Be sure that the root of the tongue is acted upon and drawn forward. Continue this action until signs of re-animation are observable, which should be the case in from ten to twenty minutes. The motion thus imparted to the tongue should be regular and rhythmical in both its tractions and relaxations. If, in attempting to seize the tongue, it is found that the jaws are closed and the teeth clenched, open them by the finger if possible, or failing this, by a wedge-shaped piece of wood, the handle of a pocket knife, or anything of the kind that may be at hand. Keep them wedged open.

The object of the traction effort on the tongue is to reinflate the lungs, and by that means re-animate the body. The rescuers' efforts should not be relaxed at the first appearance of re-animation; those efforts must be continued until there is sufficient indication that this has been secured to such an extent as to ensure regular respiration by the patient without artificial aid. The patient should then be wrapped in such clothing as can be readily got together, capable of affording warmth, and removed to the nearest hospital, infirmary, &c., or to his own residence if at hand, and placed in bed between blankets, with warm water bottles to his feet. A little brandy may be administered as soon as the patient is able to swallow, the object now being, by warmth, to promote the circulation of the blood.

Another system is as follows:—Lay the patient on his back in the open air. Remove his neckcloth and unfasten his shirt. Make a roll of clothing or anything else at hand and place it under his shoulders. The roll must be sufficiently large to so support the spine as to allow the head to fall downwards and backwards. The patient's mouth should be opened and the tongue drawn out to free the throat. The rescuer, kneeling behind the patient's head and facing him, should then grasp his elbows and draw them well over the head of the sufferer so as to bring them almost together above it, and there hold them for some two seconds. He should then carry them down to the sides and front of the chest, firmly compressing it by throwing his weight upon them. After some two seconds the action should be repeated, and continued at the rate of sixteen times per minute. Under this action the extension of the arms expands the chest walls, as in inspiration or taking breath, and if the throat is clear the air will rush into the lungs. When the arms are brought down to the sides of the chest, compressing it, the air is expelled as in expiration. The action must be regular and rhythmical, and energetically and tirelessly persisted in until the breathing of the patient has again become normal. It is possible that this may not be assured in less than an hour.

If an assistant is at hand, both systems may be employed at the same time, the tongue being drawn out as the arms are raised, and allowed to go back as the arms are depressed, the combined action being as nearly as possible perfectly uniform.

The first opportunity should be taken advantage of, by the services of any one within call, to obtain the presence of a medical man, who should, upon his arrival, be requested to take charge of the further treatment of the patient; the purpose of the foregoing instructions being to enable such restorative measures to be adopted as may be possible, pending the arrival of a qualified medical practitioner.

It is to be understood that, although there is to be no hesitation in shutting down any machinery the movement of which for the time being may endanger life, it is important that the machinery should, especially where it is employed for the lighting of railway stations, thoroughfares, buildings, &c., made use of by the public, be set in motion again as rapidly as possible. It must not be forgotten that the sudden withdrawal of light from a busy passenger or goods station, or from rooms, offices, &c., may be attended with the most serious consequences; and for this reason no time must be lost, should occasion ever arise for suspending the light, in reinstating it.

The foregoing Instructions were issued by the author for the guidance of employés of the Midland Railway, and others whom it might interest, on the 25th September, 1894.

More condensed Instructions to the same purpose, graphically describing the course of action, have also been issued by the Editors of the 'Electrical Review.'

SPECIFICATION FOR TELEGRAPH POLES.

Description.—Norwegian red fir, sound, hard grown (i.e. well hearted, and with the annular rings closely pitched), straight, free from large or dead knots and other defects, and to have the bark completely removed. To have been felled between the 1st of November, —— and the 29th of February, ——.

Each pole to contain the natural butt of the tree, which shall be sawn square after being felled.

No timber to be cut or trimmed away from the butt end so as to reduce its natural size.

The dimensions of the poles to be in accordance with the Table given at the end of this Specification, and the numbers and places of delivery to be as specified in the Schedules given on the next page.

Mode of Delivery.—All poles to be delivered carriage free, either—

(*a*) Upon the wharf of the creosoting contractor, or some other equally accessible wharf, if so directed.

(*b*) Into lighters provided for the purpose by the creosoting contractor at the nearest convenient point to his works, if it be impossible for the vessel to come alongside his wharf; or

(*c*) Into railway trucks at the place of delivery if the creosoting works are not accessible by water carriage; or

(*d*) Into waggons or timber carriages provided by the creosoting contractor to receive the poles as they are discharged from the vessel.

The poles, if floated or immersed in water after leaving the vessel in which they are imported, shall not remain so immersed longer than seven days.

Time of Delivery.—Of the total number of each length of pole under each class:—

Twenty-five per cent., or more	on or before the.................
Twenty-five per cent., or more	ditto ....
Twenty-five per cent., or more	ditto ....
Twenty-five per cent., or the residue	ditto ....

Sizes of Light Poles.				Sizes of Medium Poles.				Sizes of Stout Poles.			
Length.	Diameter at Top.		Minimum Diameter at 5 ft. from butt end.	Length.	Diameter at Top.		Minimum Diameter at 5 ft. from butt end.	Length.	Diameter at Top.		Minimum Diameter at 5 ft from butt end.
	Minimum.	Maximum.			Minimum.	Maximum.			Minimum.	Maximum.	
Feet.	Inches.	Inches.	Inches.	Feet.	Inches.	Inches.	Inches.	Feet.	Inches.	Inches.	Inches.
18	5	5¾	6	18	5½	6½	7¼	18	—	—	—
20	5	5¾	6	20	5½	6½	7½	20	—	—	—
22	5	5¾	6¼	22	5½	6¾	7¾	22	—	—	—
24	5	5¾	6½	24	5½	6¾	8	24	—	—	—
26	5	6	6¾	26	5¾	7	8¼	26	7½	9	10¼
28	5	6	7	28	5¾	7	8½	28	7½	9½	10½
30	5	6	7¼	30	6	7¼	8¾	30	7½	9½	10¾
32	5	6¼	7¼	32	6	7¼	9	32	7½	9¾	11
34	5	6¼	7½	34	6	7½	9¼	34	7½	9¾	11¼
36	5	6½	7¾	36	6	7½	9½	36	7½	9¾	11½
40	5	6½	8	40	6	7½	9¾	40	7½	9¾	12
45	5¼	6¾	8¾	45	6½	8	10¾	45	7¾	10	13
50	5¼	7	9½	50	6½	8¼	11½	50	7¾	10¼	13¾
55	5½	7½	10¼	55	7	8¾	12½	55	8	10½	14¾
60	5½	7¾	11	60	7	8¾	13¼	60	8	10½	15½

POST OFFICE SPECIFICATION FOR GALVANISED IRON WIRE.

NOTE.—In the following Specification the term "piece" shall be understood to mean a single length of wire without weld, joint, or splice of any description, either before being drawn or in the finished wire; a "coil" shall be held to mean a piece of wire in the form of a coil; a "bundle" two or more coils properly bound together; and a "parcel" shall be any quantity of manufactured wire presented for examination and testing at any one time. A "mil" is the one-thousandth part of an inch.

1. The wire shall be drawn in continuous pieces of the respective weights and diameters given in the Table hereunto annexed. Every piece may be gauged for diameter in one or more places.

2. The wire shall be made from charcoal puddled bars, shall be perfectly cylindrical, uniformly annealed, soft, pliable, free from scale, inequalities, flaws, splits, and other defects, and shall be subject to the tests hereinafter provided for.

3. The wire shall be well galvanised with zinc spelter, and this will be tested by an officer appointed by the Postmaster-General to inspect and test the wire, and hereinafter called the Inspecting Officer, taking samples from any piece or pieces and plunging them into a solution of sulphate of copper saturated at 60° Fahrenheit, and allowing them to remain in the solution for one minute, when they are to be withdrawn and wiped clean. The galvanising shall admit of this process being four times performed with each sample without there being, as there would be if the coating of zinc were too thin, any sign of a reddish deposit of metallic copper on the wire. Samples taken from pieces of the 800-lb. wire shall also bear bending round a bar $2\frac{1}{2}$ inches in diameter without any signs appearing of the zinc cracking or peeling off; the 600-lb. wire shall similarly bear bending round a bar $2\frac{1}{4}$ inches in diameter; the 450-lb. and 400-lb. wire round a bar 2 inches in diameter; and the 200-lb. wire round a bar $1\frac{1}{2}$ inch in diameter.

4. For the purpose of testing the wire as regards freedom from splits, it shall, after having been galvanised, be passed under and over four or more rollers or pulleys, placed at such distances and in such positions (*see* subjoined diagram) as the Inspecting Officer shall from time to time determine; it shall then be sufficiently stretched

or straightened to remove all bends or sinuosities by being passed round drums, either varying in diameter or differentially geared as to speed. This stretching or straightening process shall be done to the satisfaction of the Inspecting Officer.

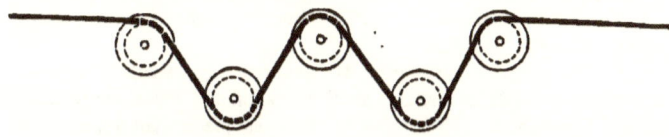

5. If, during the process of straightening, more than five per cent. of the pieces break or show any defect, the whole of the broken pieces shall be rejected. If not more than five per cent. prove defective, the whole of the broken pieces will be accepted, provided always that the wire passes all subsequent tests, and that no piece be less than 80 lb. (English avoirdupois) for the 800-lb. wire, 60 lb. for the 600-lb. wire, 40 lb. for the 450-lb. and 400-lb. wire, and 20 lb. for the 200-lb. wire. The persons tendering, herein called "the Contractors," shall not weld, join, or otherwise splice any such broken pieces as may be accepted, but shall first bind them in separate coils, and then bind such coils together to form a bundle of the standard weight, so that the broken pieces may either be jointed on the work before being paid out, or be chosen for short lengths when required.

6. Every piece may be tested for ductility and tensile strength, and five per cent. of the entire number of pieces may be cut and tested in any part. Pieces cut for this purpose, or for weighing samples, shall not be welded or jointed together again by the Contractors, but shall be treated in the same manner as the broken pieces referred to in paragraph 5.

7. To prove its ductility the wire shall be capable of bearing the number of twists set down in the Table without breaking or showing any sign of splitting or other defect. The twist-test will be made as follows:—The wire will be gripped by two vices, one of which will be made to revolve at a speed not exceeding one revolution per second. The twists thus given to the wire will be reckoned by means of an ink mark, which forms a spiral on the wire during torsion, the full number of twists to be distinctly visible between the vices, no fractions being reckoned.

8. Tests for tensile strength may be made with a lever or other machine which has the approval of the Inspecting Officer, who shall

be afforded all requisite facilities for proving the correctness of the machine. The wire shall at first lift a weight equal to at least nine-tenths ($\frac{9}{10}$ths) of the minimum tensile strength entered in the Table for the size under trial, and the remaining tenth shall be added gradually by convenient ordinary weights of not less than one-tenth ($\frac{1}{10}$th) of the remainder, i.e. one-hundredth ($\frac{1}{100}$th) of the minimum tensile strength.

9. The electrical resistance of each test-piece shall be reduced according to its diameter, and shall be calculated for a temperature of 60° Fahrenheit. The length of such test-piece shall not measure less than one-thirtieth ($\frac{1}{30}$th) part of an English statute mile. In the event of dispute as to the diameter of any test-piece, the Inspecting Officer may have such test-piece weighed, and if the weight per mile be either more or less than the standard weight, the resistance shall be not so high as that when multiplied into the weight per mile it would exceed the constant number shown in the Table, and in all cases where the product is greater than this constant, the wire will be rejected.

10. It must be understood that the tests referred to in paragraphs 3, 7, 8 and 9 are to be applied to the wire after it has been passed through the straightening process specified in paragraph 4.

11. If ten per cent. of any particular parcel of wire fail to meet all or any of the requirements of this Specification, and of the Table, the whole of such parcel shall be rejected, and on no account shall such parcel or any part thereof be again presented for examination and testing, and this stipulation shall be deemed to be, and shall be, treated as, an essential condition of the Contract.

12. Each piece, when approved by the Inspecting Officer, shall be smoothly and uniformly coiled so that the eye of the coil shall be not less than 26 inches or more than 30 inches in diameter, and each coil shall be separately bound with black varnished binders, and in no case shall two or more pieces be linked or otherwise joined together.

13. The coils shall be made up in bundles, properly bound, within the limits of weight shown in the Table. Each bundle of approved wire shall be weighed separately, and its weight (in English lb. avoirdupois) stamped on a diamond-shaped metallic label which shall be provided by the Contractors, the label being firmly affixed to the inner part of the bundle. The Contractors shall also provide the assistance necessary for properly affixing to each coil or bundle of approved wire, under the direction of the Inspecting Officer, a metallic seal which will be provided by the Postmaster-General.

TABLE referred to in the foregoing SPECIFICATION.

Weight per Mile.			Diameter.			Tests for Strength and Ductility.						Resistance per Mile of the Standard Size at 60° F.	Constant, being Standard Weight × Resistance.	Weight of each Piece* (or Coil).		Weight of each Bundle.	
Required Standard.	Allowed.		Required Standard	Allowed.		Breaking Weight.	Number of Twists in 6 in.	For Breaking Weight not less than	Number of Twists in 6 in.	For Breaking Weight not less than	Number of Twists in 6 in.						
	Min.	Max.		Min.	Max.	Min.	Min.		Min.		Min.	Max.		Min.	Max.	Min.	Max.
lb.	lb.	lb.	mils.	mils.	mils.	lb.		lb.		lb.		Stand'rd Ohms.		lb.	lb.	lb.	lb.
800	767	833	242	237	247	2,480	15	2,550	14	2,620	13	6·66	5,328	90	120	90	120
600	571	629	209	204	214	1,860	17	1,910	16	1,960	15	8·88	5,328	90	120	90	120
450	424	477	181	176	186	1,390	19	1,425	18	1,460	17	11·84	5,328	90	120	90	120
400	377	424	171	166	176	1,240	21	1,270	20	1,300	19	13·32	5,328	90	120	90	120
†400	377	424	171	166	176	1,025	23	. .	. .	1,075	20	11·84	4,736	90	120	90	120
200	190	213	121	118	125	620	30	638	28	655	26	26·64	5,328	40	·65	80	130

* Except in the case of pieces cut for testing, as provided for in paragraphs 5 and 6 of the Specification.
† This line gives the particulars for charcoal wire.

SPECIFICATION FOR CONDUCTIVITY COPPER WIRE FOR TELEGRAPH PURPOSES

(BASED ON THAT EMPLOYED
BY THE BRITISH POSTAL TELEGRAPH DEPARTMENT).

NOTE.—In this Specification the term "piece" shall be understood to mean a single length of wire without joint or splice of any description, either before being drawn or in the finished wire; a "coil" shall be held to mean a piece of wire in the form of a coil; and a "parcel" shall be any quantity of manufactured wire presented for examination and testing at any one time; a "mil" is the one-thousandth part of an inch.

1. The wire shall be drawn in continuous pieces of the respective weights and diameters given in the Table hereunto annexed, and every piece may be gauged for diameter in one or more places.

2. The wire shall be perfectly cylindrical, uniform in quality, pliable, free from scale, inequalities, flaws, splits, and other defects, and shall be subject to the tests hereinafter provided for.

3. Every piece may be tested for ductility and tensile strength, on the manufacturer's premises, or on delivery at ...., and five per cent. of the entire number of pieces may be cut and tested in any part. Pieces cut for this purpose shall not be brazed or otherwise jointed together, but each length shall be bound up into a separate coil.

4. The wire shall be capable of being wrapped in six turns round a wire of its own diameter, unwrapped, and again wrapped in six turns round a wire of its own diameter in the same direction as the first wrapping, without breaking; and shall be also capable of bearing the number of twists set down in the Table, without breaking. The twist-test will be made as follows:—

The wire will be gripped by two vices, one of which will be made to revolve at a speed not exceeding one revolution per second. The twists thus given to the wire will be reckoned by means of an ink mark which forms a spiral on the wire during torsion, the full number of twists to be visible between the vices.

5. Tests for tensile strength may be made with a lever or other machine which has the approval of the officer appointed on behalf of the ....Railway Company to inspect the wire, and hereinafter called the Inspecting Officer, who shall in this, as in all other tests

made on the Contractor's premises, be afforded by the said Contractor all requisite facilities for carrying out the tests and for proving the correctness of the machinery supplied for the purpose.

6. The electrical resistance of each test-piece shall be reduced according to its diameter, and shall be calculated for a temperature of 60° Fahr. Such test-piece shall measure not less than one-thirtieth ($\frac{1}{30}$) part of an English statute mile.

7. If, after the examination of any parcel of wire, five per cent. of such parcel fail to meet all or any of the requirements of this Specification, and of the Table, the whole of such parcel shall be rejected, and on no account shall such parcel or any part thereof be again presented for examination and testing, or for delivery; and this stipulation shall be deemed to be, and shall be treated as, an essential condition of the contract.

8. Each piece when approved by the Inspecting Officer shall be made into a coil and be separately bound; and in no case shall two or more pieces be linked or otherwise jointed together or included in any one coil. The eye of any coil shall be not less than 18 inches, nor more than 20 inches in diameter.

9. Each coil of wire shall have its weight (in English lbs. avoirdupois) stamped on a soft copper label which shall be provided by the Contractor. This label shall be firmly fixed to the inner part of the coil.

10. Each coil of wire shall be wrapped in canvas, and be delivered as required.

TABLE referred to in the foregoing SPECIFICATION.

Weight per Statute Mile.		Approx. equivalent Diameter.		Minimum Breaking Weight.	Minimum No. of Twists in 3 inches.	Maximum Resistance per mile of wire when hard, at 60° Fahr.	Minimum Weight of each piece (or coil) of wire.*
Standard.	Range Allowed.	Standard.	Range Allowed.				
lbs.	lbs.	mils.	mils.	lbs.		Ohms.	lbs.
100	97½ 102¼	79	78 80	330	30	9·10	50
150	146¼ 153¾	97	95½ 98	490	25	6·05	50
200	195 205	112	110½ 113¼	650	20	4·53	50
400	390 410	158	155½ 160¼	1300	10	2·27	50

* Except in the case of pieces cut for testing, as provided for in paragraph 3 of the Specification.

SPECIFICATION FOR GALVANISED IRON TELEGRAPH STAYING WIRE.

The Wire to be homogeneous or steely in character, uniformly annealed, smoothly galvanised, free from scale, inequalities, flaws, splits, and other defects, cylindrical in form, and of a diameter not exceeding ·176, or less than ·166 of an inch; the required standard diameter being ·171 of an inch. Each Wire to be capable of bearing a strain of not less than 1800 lbs. without breaking, with a stretch not exceeding four per cent., and to stand 12 twists in a six-inch length without fracture, and without scaling. The weight of each bundle or coil not to exceed 100 lbs. The several wires of which the strand is composed to be evenly and uniformly laid together by machinery exercising a uniform degree of tension, free from torsion, upon each wire. The maximum length of "lay" to be 12 inches, minimum 10 inches. The strand employed to be composed of 3, 5, and 7 wires. Each Wire to be without weld, joint, or splice, either in the rod before it is drawn, or in the finished wire.

SURVEY BOOK.

WORKS ORDER .... From .... To ....

No. of Pole.	Distance. Yards	Length of Pole.	Stay Rods.	Arms.		Terminal Insulators or Shackle comp.	Guards.	Remarks.
				33 in.	24 in.			

TABLET EXCHANGE BOOK.

TABLETS TAKEN FROM ....

Date.	No. of each Tablet.	Total No. of Tablets taken.	Time taken.	Station.

TABLETS RECEIVED AT ....

Signature of Signalman.	Total No. of Tablets received.	Time received.	Station.	Signature of Signalman.

WORKS ESTIMATE FORM.

.... RAILWAY TELEGRAPHS.

Telegraph Department,

.... 189

(*Reg. No.............................*)

Mr....

 Be good enough to furnish me as quickly as possible with an Estimate for carrying out the following work :—

TITLE AND DESCRIPTION.

....
....
....
....
....
....

PLAN.

WORKS ESTIMATE FORM—*continued.*

The Officer addressed is required to furnish details of the work to be done according to the following form :—

SCHEDULE OF WORK.

(Including wires to be made spare, spare wires to be brought into use, and wires recovered.)

Section.		No. of wires.	Gauge.	Open or covered work.	New or existing poles or pipes.	Length of Section.		REMARKS. (State whether to be *erected*, *recovered*, or made *spare*.)
From	To					m.	yds.	

INSTRUMENTS AND BATTERIES.

	Station.	Description of Apparatus.	Rate.	Amount.	Total.
TO BE SENT OUT.					
				£	
TO BE RECOVERED.					
				£	
				Total Debit £	
				Credit ,,	

NOTE.—The price columns will be filled in by the Superintendent's Office.

WORKS ESTIMATE FORM—*continued.*

MONEY CHARGES.

	Pay Bill.	Sundries.	Total.
Estimated cost for Salaries, Wages and Labour.			

LIST OF STORES.

Required.				To be Recovered (fit for use again).			
Description of Stores.	Quantity.	Rate.	Value.	Description of Stores.	Quantity.	Rate.	Value.
				Stores to be recovered (unfit for further use).			

TOTAL ESTIMATED COST.

Stores Required .. ..
Less Recovered .. ..
Instruments and Batteries .. ..
Money Charges .. ..

£

WORKS ORDER.

..........................RAILWAY TELEGRAPHS.
Telegraph Department,
.... 189

W.O. ....

(*Reg. No.*.....................)

Mr................................

Authority is hereby given for the execution of the following work. Be good enough to proceed with the same, so far as your duties are concerned, as early as possible; and immediately on the completion of the work, return this form, duly filled up.

TITLE AND DESCRIPTION OF WORK.

....
....
....
....

Telegraph Supt. } ....
and Engineer. }

Date............................189

PLAN.

(This plan, showing the wires erected, or brought into use, together with the instruments fixed, and their relative position to other wires, Signal Boxes, &c., is to be filled in by the Officer to whom this Order is addressed.)

to Railway Working. 315

WORKS ORDER—*continued.*

SCHEDULE OF WORK DONE.

(Including wires made spare, spare wires brought into use, and wires removed.)

Details of Work.		No. of wires.	Gauge.	Open or covered work.	New or existing poles or pipes.	Length of Section.		Remarks. (Here indicate if a wire has been made spare, or a spare wire brought into use, cut down, or erected.)
Sections.								
From	To				M.	Yds.		

APPARATUS.

FIXED.				RECOVERED.		
Office or Signal Box, &c.	Description and Number of Instruments.	Date.		Office or Signal Box, &c.	Description and Number of Instruments.	Date.

Date of Commencement..189
Date of Completion..189
Foreman in Charge of Work...
No. of P.O. Wires affected.........No. of Railway Wires affected........

Signature..

Entered in Plan by....................................
Rly. or P.O. Mr.................] „ in Works Order Book by..................
Advised as to proportion chargeable. „ in Mileage Book by..
.......................189 „ in Instrument Book by......................................
Mileage Stores, and Labour Charges, }......................
Examined by

WORKS ORDER—*continued.*

STORES AND PAY BILL CHARGES.

Labour Charges.	Stores Charges.	Total.	REMARKS.
£ s. d.	£ s. d.	£ s. d.	
Estimate.....			

ACTUAL EXPENDITURE.

Labour as per Pay Bills.		Stores Charges.		Total Cost.	REMARKS.
Week ending	£ s. d	Stock Return.	£ s. d.	£ s. d.	

WORKS ORDER—*continued.*

DETAILS OF STORES CHARGES.															
Stores Entries in Stock Return for Month of	Requisitions.	Iron Wire. No. .... Galvd.	lbs. £ s. d.	Iron Wire. No. .... Galvd.	lbs. £ s. d.	Iron Wire. No. .... Galvd.	lbs. £ s. d.							
								DEBITS.			CREDITS.		

APPARATUS EXCHANGE TICKET.

RAILWAY. TELEGRAPH DEPARTMENT.

_____ Station, _____ 189

To Mr. _____

I have sent a _____ addressed to you at _____ by the _____ Train to-day.

Signed _____

189

from _____

sent to _____

for _____

To be returned when repaired

to _____

STORES REQUISITION.

*No.*_____

_____ RAILWAY. TELEGRAPH DEPARTMENT.

*To*_____.

_____Station,_____ 189

*Please supply the undermentioned Stores for*_____*No.*_____

Title _____ *To be sent to*_____

Quantity.	Description.	Date Supplied.	How Sent.

*Signed*_____ _____

STORES DELIVERY NOTES.

TELEGRAPH DEPARTMENT. ——————— RAILWAY.
No. _____
                                                                _189
Mr. _____

   The undermentioned Stores have been addressed to
_____ at _____ on account
of _____ as per your Reqn. No. _____
                    Signature _____

Quantity.	Description.	Tons.	Cwt.	Qr.	lbs.

NOTE.—This form to be retained by the person to whom the Stores are issued.

TELEGRAPH DEPARTMENT. ——————— RAILWAY.
No. _____
                                                                _189
Mr. _____

   The undermentioned Stores have been addressed to
_____ at _____ on account
of _____ as per your Reqn. No. _____

Be good enough to satisfy yourself of their due receipt, in good condition, and return me this form, duly signed, as an acknowledgment thereof.
                    Signature _____

Quantity.	Description.	Tons.	Cwt.	Qr.	lbs.

The above Stores have been duly received in good condition.
       Signature _____
                            Date _____

REPORT ON INSTALLATION OF APPARATUS.

———— RAILWAY.

Telegraph Department,

———————————————— Section,

————————————————— 189

DEAR SIR,

Title and No. ⎫
of Order. ⎭ ————————————————

I beg to report that on the ————————————————the following Electrical Apparatus was fixed and brought into use at

———————————————————————————

———————————————————————————

Yours truly,

NOTE.—Immediately on the fixing of any Instrument, the necessary d-tails respecting the same are to be filled in upon this form, which is to be at once despatched to the Superintendent of the Department.

STORES ACCOUNTS.

DAY BOOK.

Date.	Quantity.	Description.	Requisition.	Works Order.	Delivery Note.	Name.	Station.

Y

STORES ACCOUNTS—continued.
STORES RECEIVED.
Month ending..............................

Date.	Works Order, No. and Title.	Description of Stores.	Quantity or Weight.	Rate.	Value.	Chargeable to Credit of

STORES SENT OUT.
Month ending..............................

Works Order, No. and Title.	Description of Stores.	Quantity or Weight.	Rate.	Value.	Total.	Chargeable to

STORES ACCOUNTS—*continued.*

STOCK LIST.

Description of Stores	Month ending _____ 18				Month ending _____ 18			Month ending _____ 18		
	In Stock as per last balance.	Received during the month.	Sent out during the month.	Balance in Stock.	Received during the month.	Sent out during the month.	Balance in Stock.	Received during the month.	Sent out during the month.	Balance in Stock.

INDEX.

ACCIDENTS from shock, precautionary instructions, 295-299
Acknowledgment of signals, 120, 121
Administration, 260-292
— engineering branch, 262-272
— traffic branch, 272-292
Advantages of boxing for wires, 54, 57, 58
— copper wire, 19-22
— good material and workmanship, 2, 3
— iron pole arms, 9
— Leclanché battery, 76, 77
— No. 11 wire for binding, 24
Apparatus exchange ticket, 318
Arm bolts, 28
Armoured cables, 57
Arrangement of wires, 11
Arresters, lightning, 208
Automatic block signalling, 143-154
— signal lock, Tyer's, 188, 189

BAR, expansion, for light indicator, 197-200
Batteries in use, 75-78
Battery, bichromate, 75
— — for compound working, 76
— Daniell, 76
— Dry cell, 77
— Leclanché, 75
Bell code, 114, 115
Binders for copper wire, 24, 25
Binding wire for joints, 26
— — for stays, 24
— — No. 11, 24
— — No. 16, 23

Block, Preece's one-wire, 98-102
— — three-wire, 95-98
— Pryce and Ferreira's three-wire, 107-109
— single-needle, three-wire, 88-91
— — — for single line working, 91, 92
— — — trigger lock, 90, 91
— Spagnoletti's, 93, 94
— Sykes' interlocking, 175-180
— Tyer's interlocking, 185-188
— Tyer's tablet, 84, 103-107, 126-136
Block bell, direct action, 113
— — Preece's, 109-113
— — with indicator, 110-112
— — relay, 113
— — telephone, 71-73
Block instruments, 83-85, 88-142
— — mode of fixing, 118, 119
Block section, 79, 82, 83
— signalling, 79-82
— — affirmative system, 79-82
— — automatic, 143-154
— — — electro-pneumatic, 148-150
— — — Hall system, 150-153
— — — Liverpool overhead railway, 145, 146
— — — observations on, 143, 145
— — — principles, 154
— — — Timmis' system, 145-148
— — block section, 79
— — definition of terms, 79
— — general remarks, 117-121
— — instruments for, requirements of, 83
— — interlocking, 155
— — — observations, 156-158

Block signalling, interlocking, single-needle, 158–160
— — interposition of foreign currents, 120
— — junction working, 115–117
— — length of section, 82, 83
— — limitation of number of signals, 81
— — miscellaneous appliances, 190–207
— — one-wire system, 83, 84
— — permissive system, 79
— — positive system, 79–81
— — principles of, 83
— — single line working, 122–142
— — staff and ticket, 124, 125, 141
— — — system, 122–126
— — Webb and Thompson's electric staff, 136–142
Bolts for insulators, 15, 17, 18
Boxing for insulated wires, 54
Branding poles, 4, 5
Bright's bell, 61, 62

CONSTRUCTION of line of telegraph, 1–29
— — — — list of stores, 28, 29
Contact makers, rail, 163–167, 169, 176
— — Saxby and Farmer's, 169
— — Siemens' hydrostatic, 163–167
— — signal repeaters, 191–193, 195
— — Sykes', 182
Copper tapes and binders, 25
— wire, advantages of, 19–22
— — binders for, 24, 25
— — compared with iron, 21
— — durability of, 20
— — first use of, 19
— — jointing, 53, 54
— — specification for, 306, 307
— — tapes for binding, 25
— — table for stresses, 47
Covered iron wire, 23
Creosoted and unpreserved poles, life of, 2
Creosoting poles, 5, 6

DECAY of poles, 6, 7
Disc, repeater, 195–197
Dry cell battery, 77, 78
Duplex working, 62, 63

EARTH wiring, 10
Electric light and power, 213–237, 270–272
— — attendants, trimmers, &c., useful rules for, 293, 295
— lighting, 214–237
— — arc, for goods yards, &c., 215, 216, 219, 221
— — cables, 236, 237
— — casing for wires, 230
— — compensators, 217–219
— — cost, compared with gas, 232–234
— — cost of working, 231
— — distribution of light in goods yards, &c., 219, 221, 222
— — gas plant, 234, 235
— — height, &c., of lamp pillars, 221, 222
— — illuminating power, Mr. A. P. Trotter's deductions, 224–227
— — — — Mr. W. L. Preece's deductions, 223, 224
— — incandescent, for offices, &c., 214, 215
— — lamps, 227, 229
— — plant in use on Midland railway, 217
— — position of lamp pillars, 219
— staff, Webb and Thompson's, 136–142
Electrical signals, interlocking, 155
Electro-pneumatic block signalling, 148–150
Engineering branch, administration, 262–272
— — construction works and gangs, 263–265
— — daily insulation tests, 269–270
— — duties of construction inspector, 265, 266

Index. 327

Engineering branch, duties of maintenance inspector, 265, 266
— — lineman's fault report, 270
— — — weekly report, 269
— — — linemen and assistant linemen, 264
— — maintenance, periodical inspection, 268, 269
— — normal stock, 266
— — pole diagram book, 265, 266
— — sectional stores depôts, 266
— — supervision of electric light plant, 270, 271
Expansion bar for light indicator, 197-200

FELLING and seasoning of timber for poles, 3, 4
Fixed and screw bolts for insulators, 15, 17, 18

GALVANISED iron wire, specification for, 302-305
Gauges of iron wire, 20

HALL system of block signalling, 150-153
Height of wires, 30-34
Hole digging, 38
Hydrostatic contact maker, Siemens', 163-167
Hygroscopic cement, effect of, 14

INDICATOR, Tyer's train, 202-205
Indicators, light, 197-202
Insulated wires, armoured, 55, 56
— — in signal boxes, protection of, 55
Insulators, 11-13
— cleansing, 16
— corrugated, 12-14
— D.V. No. 8, 12-14
— D.V. No. 11, 12-14
— effect of hygroscopic cement, 14
— efficiency of, 11, 12
— fixed and screw bolts, 15, 17, 18

Insulators, P. O. form, 12-14
— porcelain and earthenware, 12
— principles of construction, 12-14
— shackle cone, 18
— — — improved form, Fletcher's, 18, 19
— terminal, 16, 17
— testing, 15
— uniformity of, 12
— Z, 12-14
Instruments, block, 83-85, 88-1
— — electric interlocking, 117
— — mode of fixing, 118, 119
— — not to be tested when in use, 119
— — not to be used for messages, 92, 93
— — Preece's three-wire, 95-98
— — Pryce and Ferreira's three-wire, 107-109
— — Spagnoletti's, 93, 94
— — three-wire single needle, 88-92
— — Tyer's tablet, 84, 103-107, 126-136
— for block signalling, requirements of, 83
— telegraph, 59-63, 73-75
— telephone, 63, 64, 69-72
Intercommunication in trains in motion, 258, 259
Interlocking, block instruments, 117
— considerations thereon, 157, 158
— electrical with mechanical signals, 155
— Saxby and Farmer's, 167, 168
— single-needle block, 158-160
— Spagnoletti's, 183-185
— Sykes', 175-180
— Tyer's, 185-188
I.R. and G.P. wire, 56
Iron poles, 7, 8
— tubing for underground work, 57
— wire compared with copper, 21
— — covered, 22, 23
— — — erection of, 23
— — — West's compound, 22, 23
— — galvanised, specification for, 302-305

Index.

Iron wire, gauges of, 20
— — jointing, 54
— — table for stresses, 46

JOINTING copper wire, 53, 54
— frame, 53
— iron wire, 54
Joints, binding wire for, 26
Junction working, regulations for, 115-117

LAYING out stores, 35-37
— — wire, 49
Leading-in cups, 43, 44
— wires, 69
Leclanché batteries, 75-77
Level crossings, signal instruments for, 206, 207
Light indicators, 197-202
— indicator switch, 202
Lightning protectors, 208-212
— — plate, 209, 210
— — serrated, 209
— — vacuum tube, 211, 212
Line, construction of, 35-37
— of telegraph, construction of, 1-29
Liverpool Overhead Electric Railway, block signalling, 145, 146
Loading up material, 35, 36

MATERIAL, quality and selection of, 2
Movable bridges, protection of, 207, 208

PARALLEL wires, 64-68
Phonopore, 73-75
— induction effects, 74, 75
Pole arms, 8-10
— — arrangement of, 31
— — strengthening of, 44
— — earth wiring, 10
— — iron, advantages of, 9
— — wood, 8-10

Pole, brackets, 11
— lifter, 38
Poles, branding, 4, 5
— conversion of single to H, 39
— creosoting, 5, 6
— decay of, 6, 7
— felling and seasoning of timber for, 3, 4
— fixing stays to, 41
— H, 38, 39
— inclination of, 40
— iron, use of, 7, 8
— leading-in wires, 43
— loading and unloading, 35, 36
— pitch pine, 7
— position of, 30, 31, 33
— specification for, 4, 300, 301
— staying, 32, 39-43
— — laterally, 41
— — terminal, 42, 43
— uniformity of, 40
Precautionary instructions in relation to accidental shock, 295-299
Preece's block bell, 109-113
— single-wire block, 98-102
— three-wire block, 95-98
Pryce and Ferreira's three-wire block, 107-109
Protection of movable bridges, 207, 208
Protectors, lightning, 208
Pyrometer, 197-200

RAILWAY Clearing House standard bell code, 114, 115
Regulations for junction signalling, 115-117
Repeater disc, 195-197
Repeaters, signal, 190-195
— contact maker, 191-193, 195
Reporting train, 60, 61
Report on installation of apparatus, specimen form, 321
Revolving wires, 64-67

SAXBY and Farmer's electric slot, 172
— — — interlocking system, 167-175

Screw bolts for insulators, 15, 17, 18
Shackle cones, 18, 19
Siebrosal, 78
Siemens' hydrostatic contact maker, 163-167
Signal instruments for level crossings, 206, 207
— lock, Tyer's automatic, 188, 189
— repeaters, 190-195
— — contact makers, 195
— — principles of action, 190, 191
Single line working, 122-142
— — — staff, 122, 123
— — — staff and ticket box, 124, 125, 141
— — — staff system, 122-126
— — — tickets, 123, 124, 133, 141
— — — Tyer's tablet block, 126-136
— needle message instrument, 59, 60
Snowstorms, effect of, on wires, 21, 22
Spagnoletti's block instruments, 93, 94
— electric locking system, 183-185
Specification for copper wire, 306, 307
— for galvanised iron wire, 302-305
— for machine made staying wire, 308
— for telegraph poles, 4, 300, 301
Specimen apparatus exchange ticket, 318
— leaf of Survey Book, 309
— leaf of Tablet Exchange Book, 310
— Report on Installation of Apparatus, 321
— Stores Delivery Notes, 320
— Stores Requisition Form, 319
— Works Estimate Form, 311-313
— Works Order Form, 314-317
Staff, electric, Webb and Thompson's, 136-142
— system, 122-126
— and ticket system, 124, 125, 141
Standard bell code, 114, 115
Staples, 28
Stay blocks, 27, 28

Staying, eye bolts for, 42, 43
— poles, 32, 42, 43
— — laterally, 41
— — position of attachment, 39, 41
— use of seconds wire for, 26
Staying wire, machine stranded, 26, 27
Stay protectors, 42
— rods, 27, 28
— — thimbles for, 27, 28
— spurs, 42
Stays, attachment to poles, 41
— binding wire for, 24
— hand made, 26
— importance of, 26
Stores Account, specimen leaf from Day Book, 321
— — — — Stock List Book, 323
— — — — Stores Received Book, 322
— — — — Stores Sent Out Book, 322
— Delivery Notes, 320
— list of, 28, 29
— Requisition Form, 319
Stresses of wires, tables for, 46, 47
Survey book, 32, 309
Surveying, 30-34.
Switch for light indicator, 202
Sykes' electric lock, 180-182
— interlocking, 175-180
— rail contact maker, 182

TABLE of copper tapes and binders 25
— of stresses for copper wire, 47
— of stresses for iron wire 46
Tablet block, Tyer's, 84, 103-107, 126-136
— Exchange Book, specimen leaf, 310
Tapes for binding copper wire, 25
Telegraph instruments, 59-63, 73-75
— instruments and batteries in use, 59-78
— message code, 289-292
— poles, specification for, 300-301
Telephone block bell, 71-73
— — switch, 71

z

330   Index.

Telephone circuits, 64-68, 69-72, 287, 280
— — induction, 64-68
— — leading-in wires, 69
— — parallel wires, 64-68
— — parellelism and overhearing, 66
— — revolving wires, 64-67
— hygienic, 69-71
— instruments, 63, 64, 69-72
— — number in circuit, 63
Telephonic communication, advantages of, 63
Terminal insulators, 16, 17
— poles, staying, 42, 43
Thimbles for stay rods, 27, 28
Timmis' system of automatic block signalling, 145-148
Traffic branch, administration, 272-292
— — circuit arrangements, 272-274
— — code time, 278, 279
— — collection and examination of messages, 280
— — office check sheets, 280-284
— — reorganisation of circuits, 274, 275
— — tablet check forms, 275, 276
— — telegraph message code, 289-292
— — telephone call signals, 280
— — — circuits, 278, 280
— — train reporting, 281, 285-289
— — unnecessary use of wires, 275
— — use of prefixes, 275, 277, 278
Train indicator, Tyer's, 202-205
— lighting, 238-256
— — accumulator batteries, 243-245
— — automatic relay, 249, 255
— — auxiliary engine and dynamo, 253
— — Brush Electrical Engineering Company's machine, 254-256
— — diagram of connections, 252
— — disposition of lights in carriages, 249, 250
— — electrical coupling, 246-248
— — equalisation of distribution of current, 250-252

Train lighting, Holmes' dynamos, 241-243
— — Midland Company's experiments, 243
— — on L. B. & S. C. Railway, 238-239
— — remarks, 256
— — Stone's system, 254
— reporting, 281, 285-289
Tubing, cast-iron, 57
Tyer's interlocking block, 185-188
— signal lock, 188, 189
— tablet block, 84, 103-107, 126-136
— train indicator, 202-205

UNDERGROUND iron tubing, 57
— wires, 56
Unloading poles, 36
Useful rules for electric light attendants, trimmers, &c., 293-295

WEBB and Thompson's electric staff, 136-142
West's covered iron wire, 22, 23
Wire, binding, 23, 24, 26
— copper, advantages of, 19-22
— copper and iron, relative advantages, 21
— copper, binders for, 24, 25
— — durability of, 20
— — jointing, 53, 54
— — specification for, 306, 307
— — tapes for binding, 25
— effect of snowstorms, 21, 22
— galvanised iron, specification for, 302-305
— iron, 20
— — covered, 22, 23
— — — erection of, 23
— — jointing, 54
— laying out, 49
— staying, 26
— — machine made, 27
— — seconds wire for, 26
— — specification for, 308
— West's covered, 22, 23
Wire guards, 44

Index. 331

Wires, boxing for, 54, 57, 58
— crossing lines of railway, 34
— crossing roads, 31-34
— erection of, 44, 45
— height of, 30, 31
— insulated, armoured, 55, 56
— — in signal boxes, 55
— leading-in, 69
— parallel, 64-68
— revolving, 64 to 67

Wires, stretching, effect of, 48, 49
— table for stresses, copper, 47
— table for stresses, iron, 46
— — of sags for, 48
— tension of, 45
— underground, 56
Wiring barrow, 52-54
— tongs, 49-52
Works estimate form, 311-313
— order form, 314-317

LONDON: PRINTED BY WILLIAM CLOWES AND SONS, LIMITED,
STAMFORD STREET AND CHARING CROSS.

LIST OF MANUFACTURERS

OF

RAILWAY SIGNALLING, TELEPHONE,

LIGHTING PLANT, &c. &c.

	FOLIO
BERRY (HENRY) & CO., Leeds	3
EDISON & SWAN UNITED ELECTRIC LIGHT CO., Ltd.	5
FELTEN & GUILLEAUME (W. F. Dennis & Co.)	6
GENERAL ELECTRIC CO., Ltd.	10
HARLING (W. H.)	7
PHOSPHOR BRONZE CO., Ltd.	8
ROBEY & CO., Ltd.	4
SIEMENS BROS. & CO., Ltd.	1 & 2
WRIGHT & CO. of Southwark, Ltd.	9

SIEMENS BROS. & CO.

LIMITED,

ELECTRICAL ENGINEERS.

Iron Telegraph Poles, Insulators, Batteries, Obach Dry Cells.

Telegraph Instruments.

Electric Light Plant.

Transmission of Power by Electricity.

Insulated Wires and Cables.

Magneto and Dynamo Exploders.

Line Tools.

HEAD OFFICE: 12 QUEEN ANNE'S GATE, WESTMINSTER.

WORKS: WOOLWICH, KENT.

SIEMENS BROS. & CO.

LIMITED,

ELECTRICAL ENGINEERS.

Interlocked and Block Signalling Instruments.

Winter's Automatic Block for Single Lines.

Siemens' Patent Electrically-worked Signals and Points.

Hydrostatic Rail Contact.

Railway Gong Alarms.

BRANCHES: 21 GRAINGER STREET WEST, NEWCASTLE-ON-TYNE.
261 WEST GEORGE STREET, GLASGOW.
46 & 48 MARKET STREET, MELBOURNE.

ADVERTISEMENTS.

HENRY BERRY & CO.
LEEDS, ENGLAND
— MAKERS OF —
FIXED RIVETTERS, PORTABLE RIVETTERS, PUMPING ENGINES, PULLEY-DRIVEN PUMPS, ACCUMULATORS, &c., HYDRANTS,

STEAM ENGINES,
INGOT CRANES
TRAVELLING
      CRANES
FOUNDRY
      CRANES
HOISTS
FORGING
    PRESSES
BALING
    PRESSES
WHEEL-BOSSING
    PRESSES
FLANGING
    PRESSES
PUNCHING
    MACHINES
SHEARING
    MACHINES
BLOOM
  SHEARS
WHEEL-GLUTTING
    MACHINES
SPOKE-BENDING
    MACHINES

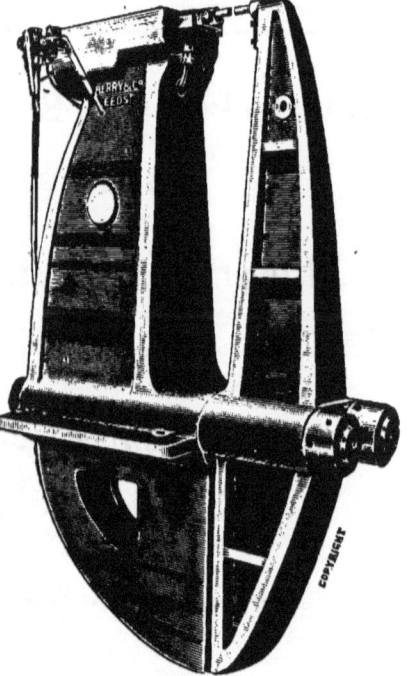

Fixed Hydraulic Rivetter (Built-up Type).

VALVES, LEATHERS, &c. &c.

HYDRAULIC MACHINERY

ROBEY & CO., LIMITED, GLOBE WORKS, LINCOLN.

Coupled Compound Horizontal Fixed Engine, fitted with Patent Trip Expansion Gear, being the simplest, most efficient and most economical of any in the market, and working practically without friction.

Open Front High Speed
Vertical Engine
for Electric Lighting.

Improved "Robey" Gas Engine.

London Offices and Show Rooms:
79 QUEEN VICTORIA STREET, E.C.

Improved Compound "Robey"
Undertype Engine.

Compound Vertical Engine for
Electric Lighting.

N.B.—ALL THE ABOVE ENGINES ARE SPECIALLY DESIGNED AND ADAPTED FOR ELECTRIC LIGHTING.

THE
EDISON & SWAN
UNITED ELECTRIC LIGHT COMPANY, LIMITED,
Ediswan Buildings, Queen Street, London, E.C.,
53 PARLIAMENT STREET, WESTMINSTER, LONDON, S.W.

Contractors to the Admiralty, Largest Mining Companies, Steamship Owners, and all principal Railway Companies.

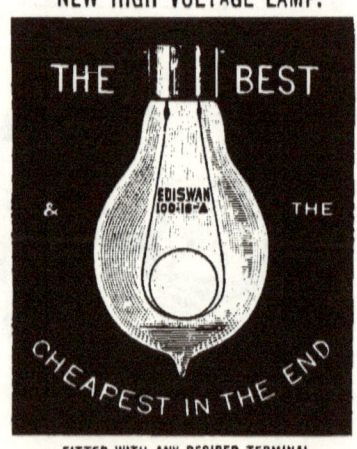

NEW HIGH VOLTAGE LAMP.

THE BEST & THE CHEAPEST IN THE END

FITTED WITH ANY DESIRED TERMINAL.

THE
WORLD-RENOWNED
EDISWAN
LAMP.

Manufacturers of everything connected with
ELECTRIC LIGHTING.

TRADE MARK—
"EDISWAN."

Fully Illustrated Catalogues free to any part of the World.

BEST ENGLISH GLASS.
ADVANTAGES:
Equality & Economy of Current Consumption.
ALL LAMPS CAREFULLY TESTED BEFORE DISPATCH.

SEE LAMPS ARE MARKED
"EDISWAN."
The other marks denote the Voltage and the
ACTUAL LIGHT EFFICIENCY.

INSIST UPON HAVING LAMPS MARKED
"EDISWAN."

FELTEN & GUILLEAUME,

Carlswerk, Mulheim-on-Rhine (Germany).

TRADE MARK.  REGISTERED.

Manufacturers of

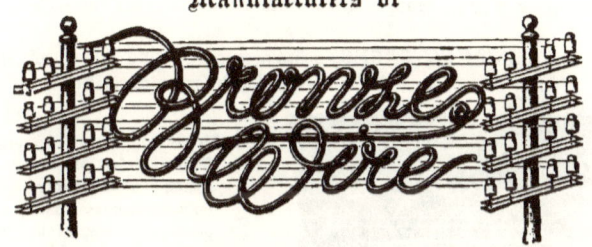

AND

HARD DRAWN COPPER WIRE
For Telegraph and Telephone Lines,

PATENT DOUBLE BRONZE AND COMPOUND WIRE,

COPPER TROLLEY WIRE FOR ELECTRICAL TRAMWAYS,

IRON & STEEL WIRE in all Qualities,

ELECTRICAL LEADS & CABLES.

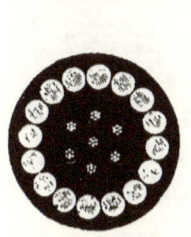  

For TELEGRAPHY,   TELEPHONY,   ELECTRIC LIGHT.

Covered and Braided Electrical Leads, Electric Light Wire, Fire and Waterproof Leads, Dynamo Wires, &c.

ALL KINDS OF WIRE ROPES.

SOLE AGENTS FOR THE UNITED KINGDOM:

W. F. DENNIS & CO., 23 Billiter St., **LONDON, E.C.**

ADVERTISEMENTS.

W. H. HARLING'S
Drawing Instruments
ARE ACKNOWLEDGED BY ALL WHO USE THEM TO BE THE
BEST IN THE MARKET.

VERY GREAT CARE is taken in their manufacture, and EVERY PIECE is examined by a competent workman before leaving the factory.

SPECIAL POCKET CASES, containing good sound instruments of my own manufacture, but not extra finish (recommended):—

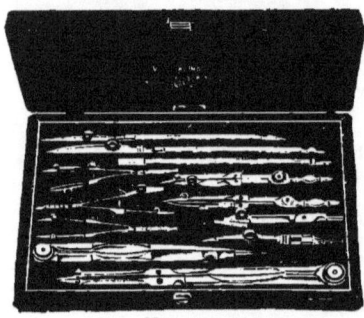

No. 721A.

721A MOROCCO CASE, containing the following electrum instruments:—6-inch double jointed, needle-pointed compass, ink and pencil points, and lengthening bar—ink and pencil double-jointed, needle-pointed bows—set of three spring bows—hair divider—two drawing pens, and ivory scale, **£3 3s.**

721B Ditto, ditto, but without set of three spring bows, **£2 10s.**

717 MAHOGANY CASE, $7\frac{3}{4} \times 5\frac{1}{4} \times 1\frac{3}{4}$, containing:—6-inch electrum pencil compass, with double knee-joints and needle point, ink point and lengthening bar—5-inch sector joint divider—one each ink and pencil needle-pointed bows—one each ink and pencil needle-pointed spring bows—two drawing pens—6-inch ivory protractor—6-inch ivory scale—two transparent set squares—and two pearwood curves, **£3.**

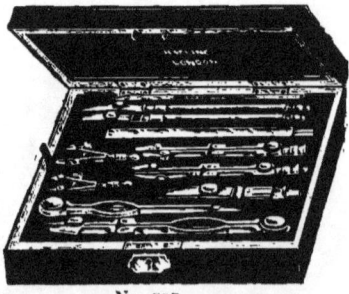

No. 717.

717A MAHOGANY CASE, as above, but fitted with plain steel points, with compass bows and spring bows, and boxwood, instead of ivory, rules, **£2 8s. 6d.**

W. H. HARLING, 47 Finsbury Pavement, London, E.C.,
Contractor to H.M. War Department, Admiralty, Council of India, &c.

ESTABLISHED 1851.   Regd. Teleg. Address, **CLINOGRAPH, LONDON.**

ILLUSTRATED CATALOGUE POST FREE ON APPLICATION.

THE
PHOSPHOR BRONZE CO., LTD.,
87 Sumner Street, Southwark,
LONDON, S.E.
And at BIRMINGHAM, MANCHESTER and ETRURIA.

Sole Makers of the Original "COG WHEEL"
and "VULCAN" Brands of

"PHOSPHOR BRONZE."

The best and most durable Alloys for Slide Valves, Bearings, Bushes, Eccentric Straps, and other parts of Machinery exposed to friction and wear, Pump Rods, Pumps, Piston Rings, Pinions, Worm Wheels, &c.

A Malleable Quality in Plates, Bars, Forgings, Stampings, Tubes, &c.

SOLE MAKERS OF
BULL'S METALS.
(MALLEABLE BRONZE.)

Strong as Steel, Malleable as Wrought Iron,
Non-corrodible as Gun Metal.

INGOTS, PROPELLERS & OTHER CASTINGS, FORGINGS, STAMPINGS, RODS & SHEETS.

Wire, Sheets, Rods, &c.,
in Phosphor Bronze, Patent Silicium Bronze, Brass, German Silver, Gun Metal and Manganese Bronze.

BABBITT METAL, "VULCAN" BRAND, PLASTIC METAL, "COG WHEEL" BRAND, "WHITE ANT" METAL cheaper than Babbitt's and equal to Antifriction Metals at much higher prices.

Please specify the manufacture of the Phosphor Bronze Company, Limited, to prevent imposition and error.

2 B

WRIGHT & CO., OF SOUTHWARK,

LIMITED,

ENGINEERS,

MILLWRIGHTS AND MACHINISTS,

157 SOUTHWARK BRIDGE ROAD,

LONDON, S.E.

PROPRIETORS AND SOLE MAKERS OF

CLARK'S PATENT AND OTHER IMPROVED

GRANITE ROLLER MACHINES

For Grinding Paints, Pigments, Inks, &c.

THE BEST AND STRONGEST MACHINES IN THE MARKET.

COLOUR MIXERS, PUG MILLS, &c.

Write for Circulars and Quotations for Paint Machinery.

REPAIRS

To Granite Roller Machines and other Machinery, Boilers and Plant promptly and efficiently carried out at Reasonable Charges, by experienced Mechanics.

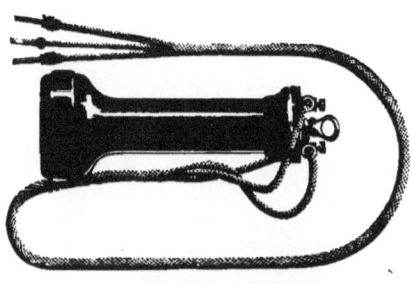

TELEPHONE RECEIVERS
In Ebonite, or Ebonite and Metal Cases.

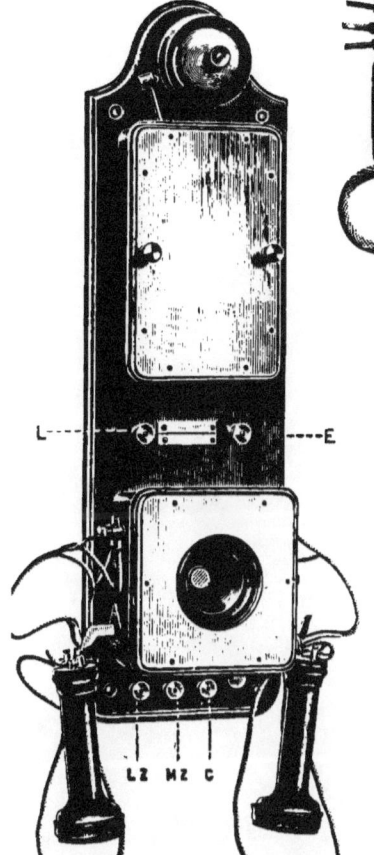

THE
"HUNNINGSCONE" TRANSMITTER
(DECKERT PATENT),
Exclusively used by the National Telephone Co., G.P.O., the leading Railways and Governments of the World.

COMPLETE TELEPHONE STATION
For Railway Signal Work.

Also Manufacturers of **DYNAMOS** and **MOTORS** on the three-phase and other systems for Railway or general purposes.

A large assortment of Telephone Stations suitable for general work always on hand.

WRITE FOR ESTIMATES.

THE GENERAL ELECTRIC CO., LTD.

LONDON—*Head Offices and Showrooms:* **69, 71 & 88 Queen Victoria Street, E.C.**
  Auxiliary Warehouses: **56 Upper Thames Street & Garlick Hill, E.C.**
MANCHESTER—*Works:* **Peel Works, Adelphi, Salford.**
BRANCHES at **45 Chapel Street, SALFORD, Manchester.**
  71 Waterloo Street, GLASGOW.
  13 Westgate Road, NEWCASTLE-ON-TYNE.
  39 Corporation Street, BIRMINGHAM.

www.ingramcontent.com/pod-product-compliance
Lightning Source LLC
Chambersburg PA
CBHW020235240426
43672CB00006B/531